Teaching Climate Change for Grades 6–12

Empowering Science Teachers to Take on the Climate Crisis Through NGSS

Kelley T. Le

Routledge
Taylor & Francis Group

NEW YORK AND LONDON

First published 2021
by Routledge
605 Third Avenue, New York, NY 10158

and by Routledge
2 Park Square, Milton Park, Abingdon, Oxon OX14 4RN

Routledge is an imprint of the Taylor & Francis Group, an informa business

Library of Congress Cataloging-in-Publication Data
Names: Le, Kelley, author.
Title: Teaching climate change for grades 6–12: empowering science teachers to take on the climate crisis through NGSS / Kelley Le.
Description: New York, NY: Routledge, 2021. |
Series: Eye on education | Includes bibliographical references.
Identifiers: LCCN 2020056994 | ISBN 9780367752354 (hardback) |
ISBN 9780367752330 (paperback) | ISBN 9781003161592 (ebook)
Subjects: LCSH: Climatic changes–Study and teaching
(Middle school)–United States. | Climatic changes–Study and teaching
(Secondary)–United States. | Next Generation Science Standards (Education)
Classification: LCC QC903 .L396 2021 | DDC 363.738/740712–dc23
LC record available at https://lccn.loc.gov/2020056994

ISBN: 978-0-367-75235-4 (hbk)
ISBN: 978-0-367-75233-0 (pbk)
ISBN: 978-1-003-16159-2 (ebk)

Typeset in Palatino
by Newgen Publishing UK

Access the companion website: www.empoweredscienceteachers.com

Teaching Climate Change for Grades 6–12

Looking to tackle climate change and climate science in your classroom? This timely and insightful book supports and enables secondary science teachers to develop effective curricula ready to meet the Next Generation Science Standards (NGSS) by grounding their instruction on the climate crisis.

Nearly one-third of the secondary science standards relate to climate science, but teachers need design and implementation support to create empowering learning experiences centered around the climate crisis. Experienced science educator, instructional coach, and educational leader Dr. Kelley T. Le offers this support, providing an overview of the teaching shifts needed for NGSS and to support climate literacy for students via urgent topics in climate science and environmental justice – from the COVID-19 pandemic to global warming, rising sea temperatures, deforestation, and mass extinction. You'll also learn how to engage the complexity of climate change by exploring social, racial, and environmental injustices stemming from the climate crisis that directly impact students.

By anchoring instruction around the climate crisis, Dr. Le offers guidance on how to empower students to be the agents of change needed in their own communities. A range of additional teacher resources are also available at www.empoweredscienceteachers.com.

Kelley T. Le, Ed.D., is currently the Director of the Science Project at UC Irvine, where she supports K-12 teachers, educational leaders, and school districts with science, engineering, and equity for justice-centered education.

Also Available from Routledge Eye On Education
(www.routledge.com/k-12)

*STEM Road Map 2.0: A Framework for Integrated STEM Education
in the Innovation Age*
Carla C. Johnson, Erin E. Peters-Burton, and Tamara J. Moore (eds.)

*Ask, Explore, Write! An Inquiry-Driven Approach to Science
and Literacy Learning*
Troy Hicks, Jeremy Hyler, and Wiline Pangle

Rigor in the 6–12 Math and Science Classroom: A Teacher Toolkit
Barabara R. Blackburn and Abbigail Armstrong

Teaching Science Thinking: Using Scientific Reasoning in the Classroom
Christopher Moore

Getting Started with STEAM: Practical Strategies for the K–8 Classroom
Billy Krakower and Meredith Martin

Inquiry-Based Science Activities in Grades 6–12: Meeting the NGSS
Patrick Brown and James Concannon

For my best friend and partner,
John Trung Le.
Thank you for being my constant and believing in me.
I love you more than all the water in the universe.

For my role model,
Jackie Tuyet-Lan Ho (Mom).
Thank you for your sacrifices, love, and strength.
My persistence and love of learning and science
comes from you.

For my two amazing kids,
Westin and Russell Le.
I know the world is better because you two are in it.
I love you both 3000.

For my educator circle,
Thank you for inspiring me to write this book.
The rippling impacts that we have on students
is tremendous. Let's continue to leave behind a
legacy grounded in love and hope.

Contents

A Pulse on the Current Climate • A Moment to Reflect
• Beginning Where You Are • Shared Leadership
• Purpose • Overview of Chapters • Learning Icons
• Trust the Process

A Story about Tradition • An Opportunity to Challenge
Science Education • Unveiling Your Teacher Disposition
• Who Said, "The Scientific Method"? • Diving Deeper into
the Process of Science • Getting Cozy with the Nature of
Science • Equity and Antiracist Science Teaching • Putting
the Pieces Together

A Story about Root-Bound Plants • A Second Look at the
NGSS for Climate Science • A Fresh Take on the NGSS
• The Role of Climate Literacy and NGSS • Climate
Change as a Socioscientific Issue • Deeper Dive into SSI
• Moving Forward with Confidence

Meet the Author

Kelley T. Le, Ed.D., is currently the Director of the Science Project at UC Irvine where she supports K-12 teachers, educational leaders, and school districts with science, engineering, and equity for justice-centered education. She has served in the educational field as a high school science teacher, instructional coach, consultant, program coordinator, and program director. Her prior roles include being the California Science Instructional Specialist where she worked closely with Title 1 as well as rural and high socioeconomic status districts. She was a former science curriculum consultant for the UCLA Curtis Center, program designer and facilitator for the UCLA Science Project, and created a High School Nanoscience course with support from the UCLA C(n)SI. She currently develops and facilitates programs in the areas of climate change education, nanoscience, next-generation science, science and equity, and mentoring.

Dr. Le graduated from Leuzinger High School in the inner city of Los Angeles, and is a first-generation student. She received her educational training from UCLA's Center X teacher education program, and the UCLA Educational Leadership Program (ELP). She is beyond appreciative for the opportunities that ELP has provided her to become a research-grounded leader bringing climate change to the forefront of education. She attributes much of her success to her resilient mother who emigrated from Vietnam as a refugee, and currently works for NASA and Boeing. Finally, she is forever grateful to the thousands of students that she has had the honor of teaching in Los Angeles and Orange County that have shaped her as a human and educational leader.

Abbreviations

AP	Anchoring Phenomenon
CCC	Crosscutting Concepts
CER	Claim, Evidence, Reasoning
CRRP	Culturally Relevant and Responsive Pedagogy
CSTA	California Science Teachers Association
CSU	California State University
DCI	Disciplinary Core Ideas
ECCLPS	Environmental and Climate Change Literacy Project Summit
EP&C	Environmental Principles and Concepts
ESS	Earth and Space Science
ETS	Engineering, Technology, and Application of Science
GCC	Global Climate Change
GHG	Greenhouse Gas
IP	Investigative Phenomenon
ISTE	International Society for Technology Education
MLL	Multilingual Learner
NAS	National Academies of Science
NASA	National Aeronautics and Space Administration
NCSE	National Center for Science Education
NGSS	Next Generation Science Standards
NOAA	National Oceanic and Atmospheric Administration
NOS	Nature of Science
NRC	National Research Council
NSTA	National Science Teachers Association
PD	Professional Development
POS	Process of Science
SEL	Social and Emotional Learning
SEP	Science and Engineering Practices
SDG	Sustainable Development Goals

SPED	Special Education
SSI	Socioscientific Issues
STEM	Science, Technology, Engineering, and Math
TFGCC	Teacher Friendly Guide to Climate Change
TOA	Theory of Action
UC	University of California
UCLA	University of California, Los Angeles
URM	Underrepresented Minority
US	United States
YPCCC	Yale Program on Climate Change Communication

Exhibits

Figures

Tables

Foreword

On Monday October 30 in 2017 I received an interesting email from NOAA Public Affairs. It was from Kelley Le, a science educator and department chairperson at the time for a Title 1 school in Los Angeles and was "extremely passionate about disseminating climate change education to as many students as possible." She was looking for and researching programs that support educators to teach climate change topics while pursuing her doctorate degree from UCLA. As the NOAA Climate Education Coordinator, I had been working to support teachers to incorporate the climate content into the classroom since 2005. Later that day I called Kelley and we began to partner to help teachers use and leverage the NGSS to increase students' scientific literacy, skills, and knowledge to effectively address the challenges and opportunities that climate change presents to communities across the nation.

Over the years, Kelley and I have explored this shared mission in numerous workshops, dialogs, and the 2019 Environmental and Climate Change Literacy Project and Summit (ECCLPS) held at UCLA. At the summit we explored with participants the alignment between the NGSS and climate change topics (using the California 2016 Science Framework) across the grade levels. In our breakout we explored how learning progresses in the NGSS related to climate change, and how to support all teachers in California. At the summit, Kelley mentioned she was working on this book and I was excited to see where her deep experience and passion would take her. Addressing the climate crisis requires transformation of our social systems, food systems, electricity production, land use, and transportation. It also requires transformation in how we teach our coming generations who need to prepare for a very different world. I think this book provides a much-needed support to help teachers reenvision their teaching as they start their journey of transformation.

Let me make the case for why this transformation of teaching is needed. Education not only enables society to benefit from the projections climate change science offers us, but also plays a central role in several reinforcing feedback loops that can accelerate the growth of climate-informed decision-making at all levels, social will and support for action, and a climate solution transition-ready workforce. Education in all its diverse forms (science, STEM, environmental, conservation, civic, place-based, etc.) needs to broadly support communities as they transition to low-carbon and resilient economies. The 2009 Climate Literacy guide, which I led for the government, approved by 18 U.S. federal science agencies and a consortium of science and education partners, concluded

> *as nations and the international community seek solutions to global climate change over the coming decades, a more comprehensive, interdisciplinary approach to climate literacy – one that includes economic and social considerations – will play a vital role in knowledgeable planning, decision making, and governance*
>
> (U.S. Global Change Research Program, 2009, p. 8; italics in original)

Teachers across the country are preparing to teach students the science and engineering called for in the new science framework. These new concepts address global challenges and opportunities posed by climate change, such as generating sufficient clean energy, building climate resilience for businesses and communities, maintaining safe supplies of food and clean water, and solving the problems of global environmental change. Consequently, the amount of time teachers spend on these issues is increasing significantly. As a result, new interdisciplinary models of education to support learners of all levels, are needed to foster climate, energy, literacy, and action. Armed with newfound knowledge and skills, students are increasingly able to contribute to and accelerate climate action in their communities.

Consider this book a call to action for you to join your fellow education leaders and teachers as we collectively transform our practices to prepare students for the climate solutions that are already underway. There are amazing and highly effective programs to support you and your students. Come explore them in this book and beyond, learn what works from others, and build your own agency and hope for the future so you can model this for your students. Even as the science becomes more dire and the time to address this challenge shortens, today's students need to be ready to accelerate the climate solutions of today and tomorrow. So many impressive transitions are already underway, from solar and wind farms to reducing food waste and moving to plant-based diets. As Paul Hawken, lead author for the book *Drawdown* said, "addressing it [climate change] is a pathway to transformation, creating a far better civilization than the one we live in now" (Hawken, 2017). We need to inspire and support our students to design a better future that solves the climate change challenge while improving our quality of life at the same time.

Today's youth are inheriting the unparalleled impacts of climate change. They are also among our most powerful champions for a sustainable, climate-resilient future. This elevates the importance of preparing today's students to implement policies and develop innovations needed to realize that future.

The education needs are significant, but they are magnified by a gap in hope when it comes to engagement on issues pertaining to climate change. Americans who have hope are more likely to engage in climate change solutions and to talk about it with their family and friends. Educators are critical messengers and facilitators of climate change education and have an opportunity to cultivate much needed hope for the future in their students. As educators, we need to know that hope is a precondition to action and these are times for action.

Over the last 15 years we have learned important things about how to close the education gap. Social science research is also clear: acquiring knowledge about climate change does not necessarily move individuals to action. Affective and social forces often influence risk perception and actions around climate change. Thus, knowledge must be paired with affect, beliefs, intentions, and motivation to enact change. The need for comprehensive, interdisciplinary climate change education is more important now than ever before.

We are privileged to live in a time when science offers meticulously observed climate trends and rigorously grounded projections of possible outcomes, offering an opportunity to learn without requiring us to experience the outcomes our science sees as possible climate futures. What we do together today will determine our shared climate future. Yet this privilege cannot be realized without building that knowledge through education. Let's do this transformational work together, it makes the heavy lift so much easier and more sustainable.

Frank Niepold
Senior Climate Education Program Manager and Coordinator,
NOAA Climate Program Office, Communication,
Education and Engagement Division

Reference

Hawken, P. (Ed.). (2017). *Drawdown: The most comprehensive plan ever proposed to reverse global warming.* New York: Penguin.

Acknowledgments

I would like to first acknowledge my family for their love and support. I am so grateful to my husband, partner, and best friend that I have known for more than half my life, John Le. I am blown away by the amount of confidence, trust, faith, and unwavering love that you continue to give. One lifetime is not long enough with you. Thank you so much for always making me feel like I can take on anything and everything. So much for just having a high school crush! To Westin and Russell Le, being your mom made me braver than I thought possible. Thank you for being my source of inspiration for writing this book to fight for the future you both deserve. Whatever you decide to pursue in life, remember that when you do find success, to turn back around and extend your hand to others. Stand up for what you believe in, love big, and know that we love you both more than all the Avengers combined. To my mom, Jackie Ho, who sacrificed everything for me to be able to have this life. I am blown away by your strength, ambition, and unconditional love. Thank you for being my mom and my first teacher. To my big sister, Katherine Nguyen, who always has my back and has always shown up for me when I needed it most. My dreams would not be possible without you to lean on. To my younger brother, Kevin Nguyen, who taught me so many life lessons. Thank you for your love and for putting up with three women in the house (I know that's not easy). To my cousin Thompson Pham, who is the coolest engineer that I admire. Thank you for instilling in me a love of nature during our "free fun time," and for showing me how to be resilient in a tough world. To my cousin Cindy Pham, who has lovingly pushed me every step of the way to find my voice. Thank you for your love, friendship, and guidance despite my crazy life's journey. Finally, thank you to my inner circle of friends and chosen family for your love and encouragement the whole way.

I am so grateful to my UCLA ELP family for their support, inspiration, and guidance. A big thank you to Cohort 24 for all the conversations, reflections, and support all those years (especially Dr. Sandy Chavez, Dr. Brooke Rios, Dr. Ces Delmuro, and Dr. James Koontz for being my support system). I am beyond grateful to have shared that space with you all and to see the remarkable work we are engaging in to disrupt and transform education. Thank you to all my professors and dissertation committee members for their support and for inspiring me to take on education for climate action.

This work was made possible by all the science teachers who came through the climate change program whose passion inspired me to write this book. Thank you for your love of teaching and for empowering students to be the agents of change we need. I also want to thank the non-profit organizations that supported my climate change educational programs. Nancy Shrodes at Heal the Bay, Katie Kozma at Reef Check Foundation, Alix Lomas at Cabrillo Marine Aquarium, and Emily Yam at Aquarium of the Pacific. Thank you for trusting me and giving up your weekend time for teachers. Your commitment to empowering communities to take action on and bring awareness to the climate crisis is so important and appreciated. Along those lines, I would also like to thank the following individuals who have either elevated my voice or provided me with support for this work: Frank Niepold (NOAA), Dr. Don Haas (PRI), Anita Davis (NASA), Dr. Brad Hoge (Formerly NCSE), Rebecca Anderson (ACE), and the many educational leaders I have gotten to know over the years. Thank you all so much for your contributions.

Thank you to my UCI colleagues and friends who continue to inspire me every day as I embark on new adventures. Dr. Acacia Warren, thank you for your friendship, support, trust, and guidance. What a privilege it is to know you and receive the light that you share with everyone in your path. Thank you to my CalTeach family (Dr. Doron Zinger, Kris Houston, and Chelsea Barilli), who work tirelessly to shape the next generation of math and science teachers to be equity-centered change agents. I am also thankful to Dr. Hosun Kang, Dr. Stephanie Reyes-Tuccio, and Dean Richard Arum for their leadership and trust. Lastly, thank you to my UCI Science Project team: Adam Woods, Allison Desfor, Gabrielle Camacho, Dr. Jennifer Cao, Jessica Lin, Juliz Ramirez, Dr. Mahya Babaie, Marianna O'Brien, Monica Maynard, Shaun Ho, and Yvonne Lopez. Thank you for moving mountains with me.

Finally, thank you to Simon Jacobs and the Routledge team for seeing the potential of what this book can be for science educators and students everywhere. Thank you for your support, advocacy, dedication, and guidance from draft to production. This book would not be possible without you all.

Introduction

Thank you for taking an interest in this book and looking for ways to integrate climate change into your curriculum. This book was written to support science educators looking to create transformative student learning experiences centered on climate change by leveraging the Next Generation Science (NGSS) Framework. As a former teacher and instructional coach, I personally understand the challenges that come with taking on the climate crisis and NGSS. This book was written as a guide to help you (1) learn more about yourself and your pedagogical decisions (because there are reasons why even the best professional development (PD) workshop doesn't transfer into the classroom sometimes), (2) provide you with practical and meaningful ways to take on both climate change and NGSS (because it's not one more thing to teach – it's *the* thing we need to teach), and (3) build your capacity to empower students through this work (because we recognize the deep yearning for climate justice and action that can no longer be ignored). We know that climate change impacts everyone, but what is often overlooked are the social, racial, and environmental injustices that are further exacerbated by this crisis. Given that the majority of people rely on the media for information on climate change (Caranto & Pitpitunge, 2015; Carter & Wiles, 2014; Hestness, McDonald, Breslyn, McGinnis, & Mouza, 2014), it is crucial for teachers to learn about the science along with effective pedagogical practices that lead to supporting students as community change agents.

A Pulse on the Current Climate

In a report published by the National Public Radio (NPR), 80 percent of parents support the teaching of global climate change in America, but 55 percent of teachers say they don't teach about climate change because it doesn't relate to their content areas (Kamenetz, 2019). Prior research also reveals that teachers who do teach about climate change, only spend between one and two hours on the topic each year (Plutzer, McCaffrey, Hannah, Rosenau, Berbeco, & Reid 2016). Furthermore, there are major teaching inconsistencies due to a myriad of factors including the need for deep climate science content knowledge, scientific literacy, or strong curriculum (Bunten & Dawson, 2014; Dawson, 2012; Hansen, 2010). What we know is that the majority of science educators do not teach about the climate crisis, so this book aims to support teachers in ways that will be meaningful to both them and their students.

Education for climate action is critical for teachers taking on this politically controversial topic to help students become informed decision-makers. It is essential to teach today's youth about the climate crisis that is quickly changing the world they are inheriting. As teachers, we hold tremendous power in the classroom to decide what is of value in that space. So how do we integrate climate science and climate change? How might we use climate change to engage students in cyclical and iterative ways of thinking to bring deeper meaning to science education? How might we move from messages of "Gloom and Doom," to helping students access their own agency as leaders who will mitigate the devastating impacts? This book will address those common questions and provide implementation support to take it back to the class. Let's begin with what you know, care about, and take action on.

A Moment to Reflect

We must start this learning journey by first identifying your intentions and realistic learning goals. The reasons for why teachers want to teach about climate change varies greatly, and identifying the lens you are wearing as you approach this book is an important first step to naming your disposition. Take a moment to consider these questions: Are you purely looking for content and/or teaching support? Are you looking for ways to empower students to take action? Are you looking for ways to engage students as critical thinkers? Are you wanting to teach science in ways that reflect the needs and demands of the twenty-first century?

I invite you to think about some reasons for why you are interested in taking on climate change, and what you're currently wondering about in regards to teaching it. Then, I encourage you to note possible expectations and goals you might have for this book, to hold yourself accountable to bringing new ideas back to your class. Finally, identify realistic challenges that may prevent you from making or implementing new ideas.

As a former science teacher and coach, I know that what teachers often learn in PD rarely makes it back to the classroom (also known as "The Problem of Enactment"). The problem of enactment stems from teachers not wanting to transfer strategies or tools that don't clearly align with their own teaching beliefs and values. As a result, educators tend to revert back to teaching the way they personally experienced or implement what has worked for them in the past. It is important to remember that teachers are people who have successfully navigated the current schooling system. So, it is not uncommon for teachers to think, "This is how I learned, so this is how you will too." If you are in that same school of thinking, I encourage you to consider the larger picture. How many people has the current educational system worked for? How do we justify knowing that underrepresented racial populations occupy only 13 percent of all STEM fields (National Science Foundation, 2020)? Is the current educational system meeting the needs of the next generation of climate warriors needed to mitigate or adapt to the impacts of climate crisis? How might climate change force us to disrupt current teaching practices? Consider your teacher disposition as you review these challenging questions and note what you are ready to explore as you step forward into growth.

Exhibit A will help to set clear intentions and learning goals to ensure you meet those goals throughout this book. As you go through the chapters, think about what risks you are willing to take as you learn ways to transform your teaching practices and curriculum.

no balance

Exhibit I.1 Setting Clear Intentions and Learning Goals Activity

1. Some of the reasons for why I want to teach about climate change are... *the world is burning up* *planet is sick*

2. I expect this book to help me... *make connections*

3. My realistic learning goals prior to digging into this book are...

4. Right now, a possible challenge to bringing these ideas back to my class are...

5. What might be possible fears or concerns I currently have in regards to teaching about climate change?

6. Do I recognize potential "risks" that might prevent me from taking this back to the class?

   a. *Risks related to my **personal** growth* (might be taking a risk by teaching new content, feeling imposter syndrome, not having enough time to plan, not familiar yet with teaching ethical dimensions of science topics, etc.).

   b. *Risks related to my **professional** growth* (might be taking a risk by having to teach this alone, trying unfamiliar resources, not having enough time to teach it, not having site support, teaching a politically controversial topic, etc.).

build background knowledge

time to evolve curriculum

There are many activities layered throughout this book that provide crucial opportunities for reflection and application of knowledge. The first step to successfully learning new material starts by unveiling our current understandings (What we currently know) and our Funds of Knowledge (How we came to know that and what cultural influences shape that knowledge). By completing the first activity, you have determined your starting point.

Beginning Where You Are

In the previous activity, I asked you to identify your disposition and learning goals so you can begin to understand the lens you might be using in your learning approach. The next activity will gauge your current beliefs around student learning experiences, curriculum design elements, and teacher attributes. Remember that a belief is an idea or assumption that we have accepted to be true. The thing about beliefs that we often forget, however, is that beliefs can change over time. As you explore this guide, you might find yourself questioning some things you believe about science education, climate change, or the NGSS. As you begin to develop different ways of thinking about science, it is important to reflect on your teaching practices and examine what would be worth bringing back to the class for students. We will revisit the following activity several times to help you track your professional growth and note any driving questions that emerge.

Tracking Your Professional Progress

Student learning experiences in my class	To what extent do I agree or disagree with this statement. What are my current beliefs or values regarding the statement? Why?
Students often discuss policies related to science.	⌀
Students have plenty of opportunities to collect and analyze scientific data or information.	not "plenty" – some collection, more analysis
Students often discuss ethical issues related to science.	not often, sometimes based on student ?s
Students engage with the Nature of Science principles.	✓
Students learn about and engage with the true Process of Science.	✓ – not formatted inquiry
Students co-construct knowledge with me every day.	✓

Student learning experiences in my class	To what extent do I agree or disagree with this statement. What are my current beliefs or values regarding the statement? Why?
Students drive the instruction in my class as capable contributors and do-ers.	✓ — I don't think they "drive" it — I choose the path + we go together

Curriculum design elements I currently value	To what extent do I agree or disagree with this statement. What are my current beliefs or values regarding the statement? Why?
I currently build all lessons/ units around anchoring or investigating phenomena.	most the phenomena exist — may need to be more cohesively planned or some units
I present climate change issues at the start of each unit or lesson.	No (!)
Students often engage in argumentation and making claims based on evidence.	✓+ everyday
Students engage meaningful discourse opportunities every day.	✓ dialogic class
My lessons/units are centered around real-world issues that are directly related to my students lives or community.	✓
Students often use media/ technology to connect classroom content to the natural world.	✓ often? sometimes

My current teacher attributes	To what extent do you agree or disagree with this statement. What are your current beliefs or values regarding the statement? Why?
I have to know everything about a particular issue before teaching about it.	— Don't have to — but like to have a strong understanding
I know everything about climate science.	Ha! No!
I feel comfortable admitting to students when I don't know the answer to their question.	✓

My current teacher attributes	To what extent do you agree or disagree with this statement. What are your current beliefs or values regarding the statement? Why?
I am comfortable teaching about open-ended issues where I cannot predict student responses.	*somethmes*
I have to feel like the expert in the room.	*Don't have to, but believe that is the expectation*
I often experience imposter syndrome even for topics that I have strong expertise in.	*✓True*
I have a strong understanding of the Nature of Science principles.	*strong?*
I have a strong understanding of the NGSS framework.	*better than some, but not as well-versed as required courses*

These reflective activities are meant to push you to acknowledge changes within your control, and potential challenges that might prevent you from bringing about that change.

Shared Leadership

As you may have already discovered, the topic of climate change is complex, emotional, and challenging. It is complex because it requires a systems-thinking approach to fully understand the scope of the problem, but teachers might organize and scaffold information in ways that make the topic seem segmented or disconnected. Climate change is highly emotional because it impacts all living things on Earth, and activating students' agency requires teachers to also address the ethical dimensions (such as social, racial, and environmental injustices). Lastly, the topic is ostensibly challenging due to its politically (not scientifically) controversial nature. To address these elements, teachers need to understand the Nature and Process of Science (also embedded in the NGSS) to develop scientific literacy in students.

Integrating climate science can have larger implications on student learning and achievement. Researchers studying educational systems in North America and abroad have found that successful schools have an inspiring vision and focus that is inclusive to teachers, staff, administrators, students, and the community (Fullan, Rincon-Gallardo, & Hargreaves, 2015). Teachers in some of the highest performing countries have adopted a mindset where they genuinely believe in their students, themselves, their colleagues, and in something larger.

> Better to light a candle than curse the darkness. There's so many people who sit back and say we're screwed, but you know what, with that one candle maybe someone else with a candle will find you, and I think that's where movements are started.
>
> (Shawn Henrichs)

There is no doubt that teaching about climate change will have a greater impact with a shared leadership approach at your school. In *Racing Extinction*, a documentary showcasing human impacts on species worldwide, Heinrichs reminds us that we are truly stronger and will go further if we mobilize on this issue together (Psihoyos 2015). Consider collaborating with a colleague that is open and willing to take on this necessary challenge. Remember the lens that you are looking through as you approach this book and note that other teachers come with different lenses. By unveiling your colleagues' teacher disposition, you can leverage their knowledge and love for teaching to join you in these efforts. Note that what you all have in common is the need to fulfill the NGSS, and climate science makes up nearly 30 percent of the framework. Think about how you might leverage the NGSS to bring your colleagues together to teach about climate change. Are there teachers in other subject areas that might have a shared vision and passion for getting the message out that you know of? Can you find allies with power to help you navigate the politics of schooling that might prevent transformation from occurring? Continue believing in your students, yourself, your colleagues, and have hope that together we can amplify our impacts to bend the curve for future generations through education.

Purpose

This book was written to provide teachers with accessible ways to integrate climate science, build their confidence around the NGSS, provide ways to activate student agency, and provide a sense of hope for the future. As a former educator, I know teachers always put students first and are willing to do what it takes to get students to succeed. My hope is that, through this book, we can connect like-minded educators together to support another. On the surface, this book is meant to serve as a guide. Underneath, I hope to push teachers' thinking about how science has been historically taught and how the NGSS could catalyze climate change learning opportunities for all. We are helping to raise the next generation of climate warriors that need to be resilient and hopeful about their futures. We have an obligation to teach about climate change to prepare students to take on the climate crisis.

Overview of Chapters

Chapter 1, "Reenvisioning Science Teaching" is useful for teachers that are ready to examine their pedagogical practices through the lens of NGSS. Starting from where you are, this chapter asks you to reflect on how much of a paradigm shift you are ready to engage in as you take on climate science for transformational teaching.

Chapter 2, "Leveraging the NGSS for Climate Change" examines NGSS-aligned instruction and the crucial role of climate change in the framework. This chapter also provides a comprehensive overview of research and data behind the frameworks grounding this book, and how these approaches have proven to be effective for both student and teacher learning. There will be many opportunities to transfer what you are learning into the classroom to experience what an NGSS lesson sounds, looks, and feels like with students.

Chapter 3, "Climate Change Is Complex, Where Do I Start?" goes over the three fundamentals of climate science as outlined by national and local educational directors from various reputable organizations. This chapter also highlights what researchers in the field of climate science education have found to be effective in teaching about climate change.

Chapter 4, "Climate Change as the Anchor" explores what it looks like to center science instruction around climate science to empower students as agents of change. Using climate change phenomena as an anchor in NGSS storylines, teachers receive practical tools and resources to develop students as critical consumers and thinkers in a digital age.

Chapter 5, "Planning and Teaching for Transformation" provides a comprehensive look at lesson planning NGSS storylines so that you can bring it back to the class. Teachers will also explore the ethical dimensions of climate change with students through a culturally relevant and responsive approach.

Chapter 6, "Education for Climate Action" provides additional support for teachers looking to transform their practices and pedagogy as equity and justice-centered educators. This final chapter explores a myriad of vetted teaching resources aligned to NGSS, while further tackling the ethical dimensions of the climate crisis to incorporate ways for students to be change agents in their own communities.

Frameworks for Successful Teaching and Learning

There are three frameworks referenced often throughout the book to support the teaching and learning of climate change content through NGSS. See Table I.1 for the frameworks and their descriptions.

The appendices at the end of the book includes referenced tools, resources, and samples of student work. Visit www.empoweredscienceteachers.com to find out more about the author, access teacher resources, and get information on current professional development opportunities.

TABLE I.1 Major Frameworks and Descriptions

Framework	Description
Culturally Relevant and Responsive Pedagogy (CRRP)	This framework recognizes the value of students' cultural backgrounds, funds of knowledge, and lived experiences in education. Ladson-Billings (1995) stresses the need for educators to tend to student achievement and learning, cultural competence, and sociopolitical awareness/critical consciousness to empower students.
Next Generation Science Standards (NGSS)	Released in 2013, this science framework includes three-dimensional learning and mastery of performance expectations. There is also an emphasis on engineering practices, climate change content, and the nature of science.
Socioscientific Issues Framework	This framework focuses on science topics that are complicated, open-ended, and controversial without straightforward solutions (such as climate change). It also looks at ways to address climate change effectively for both teachers and students.

Learning Icons

Throughout the book you will see the following icons displayed at specific points. These icons highlight opportunities to understand (1) where you are in your own practices, (2) when to connect with the network, and (3) ideal moments to try out what you learned or created with your class.

(1) Learn More About Myself   (2) Connect with the Network   (3) Take it Back to the Class

Trust the Process

"Choose courage over comfort" and remember that change takes time. It is essential to have small ambitious goals as you take on climate change, but remember to have patience for yourself and to consistently return back to those goals to reflect on your progress. The NGSS pushes teachers to model the true Nature and Process of Science for students, and although climate change allows for teachers to do that more easily, the topic has such a large scale that it is impossible to know everything about it. Thank you for trusting the process, committing the time, and positioning yourself as a learner as you take on this important challenge.

References

Bunten, R., & Dawson, V. (2014). Teaching climate change science in senior secondary school: Issues, barriers and opportunities. *Teaching Science*, 60(1), 10.

Caranto, B. F., & Pitpitunge, A. D. (2015). Students' knowledge on climate change: Implications on interdisciplinary learning. In *Biology Education and Research in a Changing Planet* (pp. 21–30). Singapore: Springer.

Carter, B. E., & Wiles, J. R. (2014). Scientific consensus and social controversy: Exploring relationships between students' conceptions of the nature of science, biological evolution, and global climate change. *Evolution: Education and Outreach*, 7(1), 6.

Dawson, V. (2012). Science teachers' perspectives about climate change. *Teaching Science*, 58(3), 8–13.

Fullan, M., Rincon-Gallardo, S., & Hargreaves, A. (2015). Professional capital as accountability. *Education Policy Analysis Archives*, 23(15). http://dx.doi.org/10.14507/epaa.v23.1998.

Hansen, P. J. K. (2010). Knowledge about the greenhouse effect and the effects of the ozone layer among Norwegian pupils finishing compulsory education in 1989, 1993, and 2005 – What now?. *International Journal of Science Education*, 32(3), 397–419.

Hestness, E., McDonald, R. C., Breslyn, W., McGinnis, J. R., & Mouza, C. (2014). Science teacher professional development in climate change education informed by the next generation science standards. *Journal of Geoscience Education*, 62(3), 319–329.

Kamenetz, A. (2019, April 22). *Most teachers don't teach climate change; 4 In 5 parents wish they did.* Retrieved May 25, 2020, from www.npr.org/2019/04/22/714262267/most-teachers-dont-teach-climate-change-4-in-5-parents-wish-they-did

Ladson-Billings, G. (1995). Toward a theory of culturally relevant pedagogy. *American Educational Research Journal*, 32(3), 465–491. https://doi.org/10.3102/00028312032003465

Plutzer, E., McCaffrey, M., Hannah, A. L., Rosenau, J., Berbeco, M., & Reid, A. H. (2016). Climate confusion among US teachers. *Science*, 351(6274), 664–665.

Psihoyos, L. (2015). *Racing Extinction*. New York: Abramorama.

National Science Board, National Science Foundation (2020). *Science and engineering indicators 2020: The state of U.S. science and engineering.* NSB-2020-1. Alexandria, VA. Available at https://ncses.nsf.gov/pubs/nsb20201/.

Part 1

Looking Back to Move Forward

1

Reenvisioning Science Teaching

Read this when:

- ◆ *You're ready to think about your own pedagogical practices (ways of teaching) to determine where you are as we align to the NGSS.*
- ◆ *You're ready to engage in a paradigm shift (ways of thinking) to transform your teaching practices.*

A Story about Tradition

It was New Year's Eve, and Audrey was given the honor of cooking her mother's famous honey ham for their big family feast. Audrey sent her partner to the store to purchase the meat and made the special request for the butcher to cut the ends off the ham. When her partner returned with the groceries, he asked about the special request to which Audrey responded, "It's how my mom does it." That night they gathered at her mother's home to celebrate, and Audrey's curiosity took hold. At the dining table where the entire family convened, she asked her mother why it was so important to cut the ends off of the ham. Thinking it had to do with how the ham needed to cook more evenly, she was surprised by her mother's response, "It's how Grandma always made it." Not entirely satisfied with her mother's response, she decided to ask her grandmother directly. After a few seconds,

Grandma responded, "I had to cut the sides off so the ham could fit into my roaster."

When you think about this story, it parallels with how many people still think about education and teaching today. Consider how much (or little) education has evolved over the decades compared to technology, transportation, or even nature. Although there is a vast amount of research detailing how students effectively learn, most of those findings rarely make it into the classroom. I encourage you to think about how some components of teaching fall in line with educational tradition and compare it to research-based teaching practices on how students learn. When you think about aligning your practices and curriculum to the NGSS, it's important to identify what can and cannot be changed about teaching. You might not have the power to change the overall structure of education yet (i.e. bell schedule, instructional minutes, academic calendar, student–teacher ratio, etc.), but you do have the power to change your beliefs about how and what should be taught. It begins by asking the right questions about teaching and learning, and then doing our due diligence to see what credible researchers have discovered about your wonderings. I invite you to complete Exhibit 1.1 to start unveiling your underlying beliefs about teaching and learning.

 As you begin to think about your current teaching practices in relation to climate science and NGSS, start the process by being honest with your starting point. Think about realistic learning goals you might have to enhance your instruction along with actionable steps to accomplish those goals. Consider writing these goals in the margin to see if they change throughout this book.

Exhibit 1.1 Common Questions about Schooling

Read through the following questions and check the boxes for ones you're most curious about.

◆ Why are students trained to raise their hands to speak? What does that teach them?
◆ How has education changed (or not) in the last century?
◆ What have researchers uncovered about how students effectively learn science?
◆ Why is there a need for the NGSS?
◆ What are the differences between NGSS-aligned curriculum and NGSS-aligned teaching? How might a teacher have one, but not the other?
◆ What might it look like to co-construct meaningful science experiences with students?
◆ Who decides what and/or how content should be taught?
◆ Why is there an emphasis on the climate crisis now in the curriculum?
◆ Is there a need for culturally relevant or responsive teaching? How confident are you with this framework?
◆ Education was initially created with only one audience in mind prior to the 1960s. Has it (if at all) transformed since to support every student?
◆ Why might it be important for students to see themselves in my science lessons? How confident are you with what this means?
◆ When do you tend to equity or justice-centered instruction? Is that one of your priorities?

 What are some things you're wondering about as you think about traditional education and the paradigm shift needed for the NGSS?

1.
2.

An Opportunity to Challenge Science Education

Don Duggan Haas, the Director of Teacher Programs at the Paleontology Research Institute (PRI), affirms that "the most valuable things in our educational system are the human resources, especially the wisdom and passion of the teaching force" (2020). In his Science in the Virtual Pub segment, he breaks down the components of the traditional educational system as a call to action to transform schooling. Haas (2020) argues that every part of society has evolved including the technology and innovations for communication, entertainment, energy capturing systems, and transportation systems, among others. Surprisingly, the one system that has not seen improvements in the last 40 years is education. Despite the vast body of research that highlights what teachers and students need to learn more effectively, not much has transferred to the classroom.

Haas (2020) notes that real problems that people care about are highly intersectional, such as the climate crisis happening all around us. What is needed in order to mitigate the impacts will be a transformation of how we think, feel, and act in our daily lives. That is not an easy ask. Furthermore, Haas contends that people tend to adopt innovations, "When they are different enough to make a difference, but not too different that you don't understand it" (2020). Essentially, teachers tend to adopt new practices that are not too far off from what they have already been doing that may yield better results. Teachers also tend to adopt new practices that are aligned with their underlying core values of what they believe reflects good teaching and learning. In order for you to invest more time into changing your practices (which will also reflect in your curriculum design), you need to know the rationale behind those ideas, and how they will yield results with ongoing classroom support. Complete Haas' adapted activity in Exhibit 1.2 to unveil more of your underlying beliefs about schooling.

Exhibit 1.2 What a Great Idea!

I want you to stop for a second, and try to forget everything you know about schooling. Imagine that a friend approaches you with a new and innovative business idea. He pitches the following:

> Hey, I have a great idea! Let's put 1000 teenagers in a building and sort them into groups of 40. Have someone talk at them for 55–90 minutes about the atom. Then move them down the hall and have someone else talk at them or engage them in activities for exactly the same amount of time about math (or another other subject). Then move them down the hall and repeat this four more times in the same day. Let's do this over and over again. Day after day. Month after month. Year after year, for centuries to come. Isn't that a great idea?

Haas exposes the structural components of our educational system that have not changed in light of research-based studies on how students learn.

Think about and/or write some thoughts you have for the following questions.

1. Using research on how students learn, can/should we redesign learning experiences to reflect those ideas? Why or why not?
2. Has this educational system done an effective job in supporting all students? Which students? To what extent? Where could we improve? Would you confidently say that schooling is 100 percent accessible to 100 percent of the students?
3. What is within your realm of control as you think about changes that can be made to improve students' learning experiences?
4. What is or will be a challenge for you as you think about aligning more to NGSS?

Unveiling Your Teacher Disposition

In order to meet you where you are, you need to first unveil your own ideas and core values about good science teaching. As stated previously, teachers make decisions based on beliefs they hold about good science instruction and how students learn. When you take action on those beliefs that is known as your theory of action (TOA). Just like beliefs, your TOA may change over time as your beliefs change over time. You can see this in action by observing another colleague teach. During that observation, you will immediately notice that what works for your colleague, might not work for you. Similarly, what is acceptable in your class, might not be acceptable in your colleague's class. When you think about it that way, you begin to understand that your actions are based on underlying core teaching values that may differ from others, which is why they might have a different approach to a similar problem. I invite you to complete Exhibit 1.3 to better understand where you currently are with your instructional practices and NGSS.

Exhibit 1.3 Aligning Teaching Practices to the NGSS

 Let's start with how you currently teach and the pedagogical decisions you make. *Think about the best science lesson that you have ever taught*. Consider the following questions to set the stage.

1. How did that lesson make you feel?
2. What were your students doing during the lesson?
3. What did your class sound like?
4. If an observer saw your class, what would they see during that lesson?
5. What are students capable of during this lesson?
6. Were there any challenges present during this lesson?
7. What was the overall structure (flow and sequence) of the lesson?

8. What are some components of a science lesson that are considered non-negotiables to you?

9. What parts and/or components of this lesson are strongly aligned to the NGSS?

10. When you think of this lesson, do you find yourself looking at the NGSS for components that you addressed in some way (like checking off specific NGSS dimensions or parts of a Performance Expectation)? Why or why not?

 Now that you have completed the above, consider the following questions to see your possible areas for growth and alignment towards the framework.

☐ Where do students see themselves in this lesson (specifically looking at their beliefs, culture, identity, interests, lived experiences, or community)?

☐ Are there any elements of the Nature of Science (NOS) present in your favorite lesson? If so, which NOS principles (if any)?

☐ What level of questioning was provided consistently throughout the lesson? Who was posing these questions? Why?

☐ How does this lesson build on the lessons prior (making it cyclical)?

☐ How will it explicitly build into future lessons (making it iterative)?

☐ What are the student discourse opportunities provided consistently?

☐ At which points do students get to teach you something?

☐ In what ways was this lesson driven by students' ideas or interest by design?

☐ How was the level of productive engagement measured?

☐ How were students positioned as scientists and capable contributors?

☐ In what ways did this lesson help develop students' scientific identity?

☐ In what ways was this lesson culturally relevant and responsive to the population you teach?

☐ How did this lesson promote social, racial, and/or environmental justice to help students access their agency to enact change in their own communities?

 Check the box to questions you are most interested in exploring. Next, write some things you are wondering about now to push your thinking about science teaching.

Who Said, "The Scientific Method"?

Moving towards the new framework, we have to address the issue of The Scientific Method. Note that most teachers didn't receive educational training explicitly on the Nature of Science through their teacher education programs or college experience, so they might be communicating about the scientific process as something that happens in a linear fashion. This is otherwise known as, "The Scientific Method." The Scientific Method essentially lays out the process of science as an organized step-by-step instructional guide. The problem with this method is that it does not align with what scientists actually do or how they engage in scientific research. Science is about engaging in practices, doing science, and understanding the diverse range of approaches one can take to understand or solve a problem. The Scientific Method does not position students as scientists by allowing them to make claims from evidence, modify claims/questions in light of new information, or engage in discourse for sense-making. The climate crisis and environmental issues that impact students' lives require complex and innovative ways of thinking. It's time to close the book on The Scientific Method and teach students that science is something you do by engaging in a variety of practices through a variety of ways. Exhibit 1.4 highlights issues with The Scientific Method, and the need to emphasize the rich nature and process of science. Remember, the shift towards the NGSS requires a shift in thinking about how we teach and position students in the classroom.

Exhibit 1.4 A Better Understanding of the Process of Science

Many students are taught that science is about proving what we already know to be true. Contrary to that idea, science is actually about discovering new concepts, ideas, or solutions in creative ways. Key concepts of the process of science, however, are rarely communicated in classrooms alongside science content and what has already been discovered. Students should be consistently asking questions and when presented with new evidence or information, ask even more questions.

 Think back to your best lab experiment, activity, or science lesson. Consider the following questions to continue exploring your TOA.

☐ When did you learn about "The Scientific Method," and do you think that process reflects how scientists operate today?

☐ When thinking about your own classroom experiments or activities, how much of the process was driven by students' interests and ideas?

☐ Are you thinking of making any changes to the lab, activity, or lesson to better align to the true process of science? If so, what support do you need?

☐ Are the modifications (if any) to your lab, activity, or lesson do-able? What challenges might you encounter as you align the curriculum more to what NGSS calls for?

☐ Are you confident in your knowledge of how the process of science is folded into the NGSS? If so, take a moment and think of how it's specifically reflected and why.

☐ How do you think the process of science builds students' scientific literacy skills for NGSS?

□ Do you think that scientific literacy is an essential skill that students should have? How might this skill be beneficial for them?

□ Are your students co-constructing information with you and/or positioned to do so?

 Check the box to questions you are most interested in exploring. Next, write some things you are wondering about as you reflect on your best science lesson.

To better visualize the process of science, UC Berkeley created an interactive tool that allows students to make their thinking visible as they explore science issues (see Figure 1.1). Looking at the visual, teachers should note how complex the process can be (in positive ways), how there are multiple entry points for students, and the variety of ways students can tap into their cultural wealth to explore different issues. When thinking through ways to make your lesson culturally relevant and responsive, you can use this tool to give your students more voice and choice with a central topic to investigate (see Appendix A on how to use the interactive tool). Exhibit 1.5 provides ideas on how to implement this tool with your students.

How science works

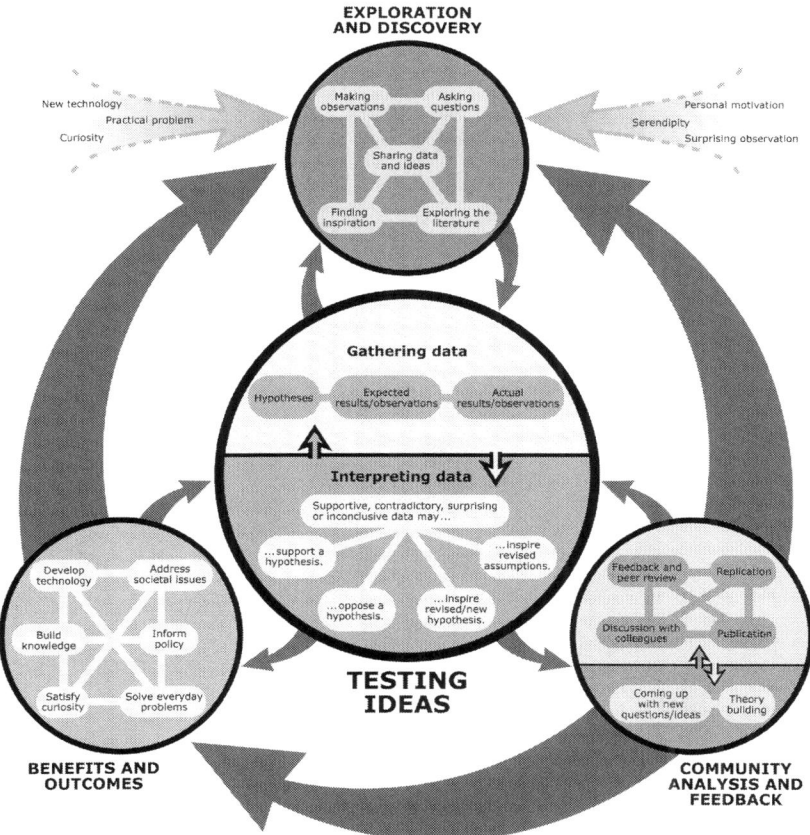

FIGURE 1.1 UC Berkeley's Process of Science Tool
Image credit: Understanding Science by the University of California Museum of Paleontology.

Exhibit 1.5 What Does This Look Like in the Classroom?

Teachers don't need to wait for the perfect moment to teach about the nature and process of science. This could take place on a shortened day, or the first week of school as you are setting classroom norms/procedures/lab safety. Revisit the tool often to showcase the variety of methods students can employ to engage in science. Allow for students to talk about their maps to engage them and the class in sense-making. This will allow students to see each other as sources of information and capable contributors to science.

1. Have students think about and identify an environmental problem they notice *in their neighborhood* that they think science could help investigate or create a solution for.
   ✓ *Example – Students view air quality data after learning about the chemical composition of air pollution. The students notice that different neighborhoods have different levels of clean air and are wondering why there are such drastic differences. Have students use the online tool in groups of three to think about how they might investigate this.*

2. Have students think about and identify a problem they notice *in their school* that they think science could help investigate or create a solution for.
   ✓ *Example – There's a rumor that the drinking water pipeline at the school is the same water that comes out of the science laboratory sinks. Students were advised to never drink water from the classroom faucet, but some students and staff are starting to wonder if the drinking fountains are safe as well. Have students use the online tool in small groups to think about how they might investigate this.*

3. Present a phenomenon and ask students to use the tool to map out how they might approach learning more about the phenomenon.
   ✓ *Example – Students watch a short 20-second clip of coral bleaching occurring in New Caledonia. The coral begins*

to change into neon blue, purple, and bright green colors rather than a pale white. Have students use the online tool in small groups to think about how they might investigate this phenomenon. Request that they chart out what questions they have about what they are seeing to help focus their investigation (allowing for different groups to pursue different interests).

4. Have students keep track of how they engaged in science for a lab experiment they are conducting.

   ✓ Example – Students just learned about the lack of access to clean water to nearly 1 million people on the planet and are exploring new water filtration methods to help address the problem. Students conduct a water filtration lab using nanofilters to remove three major contaminants (chemicals, urine, and bacteria). They compare results of water samples when carbon or zeolite are added to the nanofilters. Students are asked to bring a random sample of dirty water into class and have one chance to determine the best filtration method for their own sample. Have students use the online tool with their groups to think about how they might investigate this problem. Request that they chart out what questions they have about what they are thinking about to help focus their investigation for sense-making.

Diving Deeper into the Process of Science

As we continue to think about ways to challenge the current model of science teaching to align to NGSS, what will be needed is a paradigm shift between the old and new frameworks. The NGSS cannot be approached in the same way that teachers approached the old state standards. The old framework was essentially a checklist of concepts that teachers needed to address prior to the state exam. Rote memorization and constant recall of knowledge was valued and rewarded. In the old framework, students were positioned as sponges who absorbed information that rarely connected larger ideas together. It's also important to note that the old framework ascribed to The Scientific Method, where specific steps were provided and often there was only one predetermined answer at the end of the experience to gauge mastery. Designing culturally relevant and responsive lessons, unveiling students' current understandings, engaging in discourse, utilizing evidence to support claims, and much of the process of science was often negated with the old framework. If any of the new components took place, it was because the teacher invested personal time to learn more, and had a strong desire to integrate them into the classroom reflecting their TOA.

The NGSS is unlike any type of science instruction that we might be accustomed to or have personally experienced. The new framework emphasizes engaging in science and engineering practices, cyclical and iterative ways of learning across grade levels and within content areas, stresses the need for climate education, and pushes for equity and justice-oriented ways of teaching. This is the moment for teachers to reenvision science education for what it could be. This directly connects back to teachers' beliefs about teaching and learning because you have the power to elevate certain principles and practices or completely ignore them regardless of the evidence.

It is clear that a paradigm shift can be challenging and difficult because teachers have to reflect on deeper levels first to make the shift. Consider completing the following activity in Exhibit 1.6 to push your thinking and TOA based on what you've learned so far.

Exhibit 1.6 All Students, All Science?

What might be worth considering here is whether or not traditional approaches to schooling supports every student. Considering the following questions:

☐ Who has benefited from the traditional methods of science instruction (old framework)?

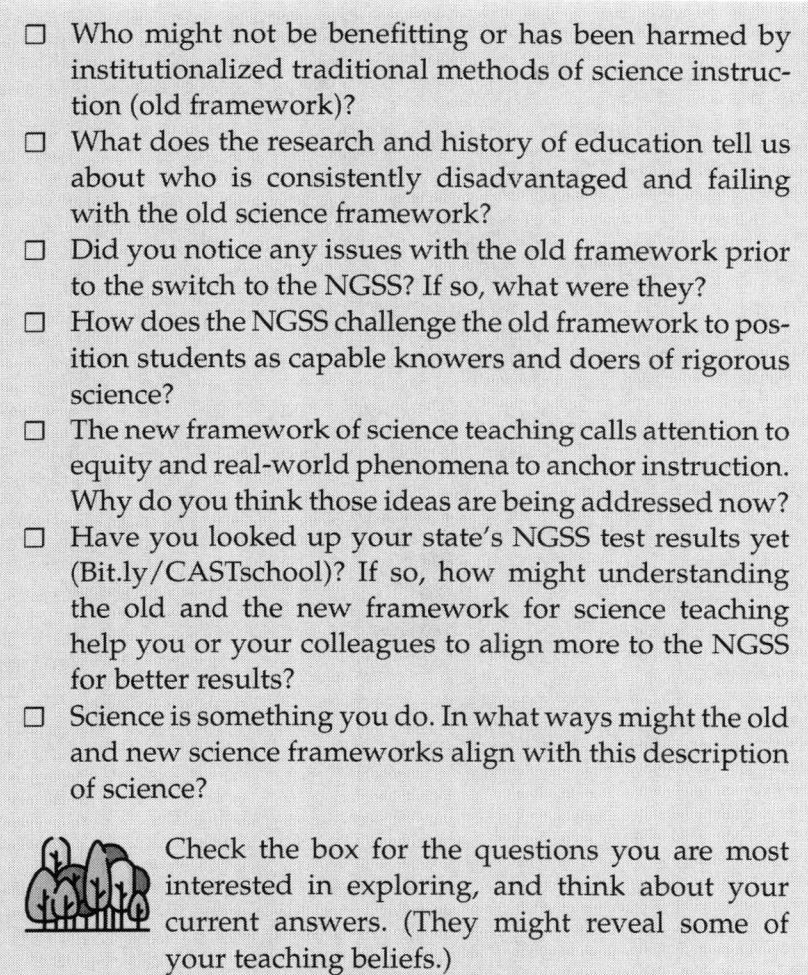

☐ Who might not be benefitting or has been harmed by institutionalized traditional methods of science instruction (old framework)?

☐ What does the research and history of education tell us about who is consistently disadvantaged and failing with the old science framework?

☐ Did you notice any issues with the old framework prior to the switch to the NGSS? If so, what were they?

☐ How does the NGSS challenge the old framework to position students as capable knowers and doers of rigorous science?

☐ The new framework of science teaching calls attention to equity and real-world phenomena to anchor instruction. Why do you think those ideas are being addressed now?

☐ Have you looked up your state's NGSS test results yet (Bit.ly/CASTschool)? If so, how might understanding the old and the new framework for science teaching help you or your colleagues to align more to the NGSS for better results?

☐ Science is something you do. In what ways might the old and new science frameworks align with this description of science?

Check the box for the questions you are most interested in exploring, and think about your current answers. (They might reveal some of your teaching beliefs.)

Getting Cozy with the Nature of Science

The NGSS highly promotes scientific literacy among students through essential topics such as climate change (NGSS Lead States, 2013). To support scientific literacy, the standards call for three-dimensional learning and push students to understand the Nature of Science (NOS). People need to be scientifically literate because they influence the implementation or formation of public science policies that may directly impact their lives. Individuals should be able to find, make sense of, or use information on topics of discussion to make well-informed decisions regarding science or technology related policies (Miller, 2016). Although 52 percent of Americans take interest in science or technology related issues, roughly 28 percent of the overall population is scientifically literate (Miller, 2016). Teachers are no doubt among this population, but accountability to the NGSS requires that many need to quickly become scientifically literate to effectively support students. The first step is to look at major NOS themes called out in the framework.

The National Research Council (NRC) and the National Science Teachers Association (NSTA) both stress that a fundamental goal for science education is to produce scientifically literate people who understand the NOS (NRC, 2013; NSTA, 2016). There are eight major principles embedded within the NGSS to help students develop skills as informed critical thinkers and decision makers. Table 1.1 highlights different NOS principles and how each relates to the NGSS Science and Engineering Practices or Crosscutting Concepts (view full table online by looking up NGSS Appendix H).

TABLE 1.1 Understandings about the Nature of Science* (NRC, 2013)

NOS Principles	Middle School	High School
Scientific Investigations Use a Variety of Methods (SEPs)	• Science investigations use a variety of methods and tools to make measurements and observations. • Science investigations are guided by a set of values to ensure accuracy of measurements, observations, and objectivity of findings. • Science depends on evaluating proposed explanations.	• Science investigations use diverse methods and do not always use the same set of procedures to obtain data. • Scientific inquiry is characterized by a common set of values that include logical thinking, precision, open-mindedness, objectivity, skepticism, replicability of results, and honest and ethical reporting of findings.

NOS Principles	Middle School	High School
Scientific Knowledge is Based on Empirical Evidence (SEPs)	• Science knowledge is based upon logical and conceptual connections between evidence and explanations. • Science disciplines share common rules of obtaining and evaluating empirical evidence.	• Science knowledge is based on empirical evidence. • Science includes the process of coordinating patterns of evidence with current theory. • Science arguments are strengthened by multiple lines of evidence supporting a single explanation.
Scientific Knowledge is Open to Revision in Light of New Evidence (SEPs)	• Scientific explanations are subject to revision and improvement in light of new evidence. • Science findings are frequently revised and/or reinterpreted based on new evidence.	• Scientific explanations can be probabilistic. • Most scientific knowledge is quite durable but is, in principle, subject to change based on new evidence and/or reinterpretation of existing evidence. • Scientific argumentation is a mode of logical discourse used to clarify the strength of relationships between ideas and evidence that may result in revision of an explanation.
Science Models, Laws, Mechanisms, and Theories Explain Natural Phenomena (SEPs)	• Theories are explanations for observable phenomena. • Science theories are based on a body of evidence developed over time. • A hypothesis is used by scientists as an idea that may contribute important new knowledge for the evaluation of a scientific theory.	• A scientific theory is a substantiated explanation of some aspect of the natural world, based on a body of facts that has been repeatedly confirmed through observation and experiment, and the science community validates each theory before it is accepted. If new evidence is discovered that the theory does not accommodate, the theory is generally modified in light of this new evidence.

(continued)

TABLE 1.1 *(continued)*

NOS Principles	Middle School	High School
		• Models, mechanisms, and explanations collectively serve as tools in the development of a scientific theory. • Laws are statements or descriptions of the relationships among observable phenomena.
Science is a Way of Knowing (CCCs)	• Science is both a body of knowledge and the processes and practices used to add to that body of knowledge. • Science knowledge is cumulative and many people, from many generations and nations, have contributed to science knowledge. • Science is a way of knowing used by many people, not just scientists.	• Science is both a body of knowledge that represents a current understanding of natural systems and the processes used to refine, elaborate, revise, and extend this knowledge. • Science is a unique way of knowing and there are other ways of knowing. • Science knowledge has a history that includes the refinement of, and changes to, theories, ideas, and beliefs over time.
Scientific Knowledge Assumes an Order and Consistency in Natural Systems (CCCs)	• Science assumes that objects and events in natural systems occur in consistent patterns that are understandable through measurement and observation. • Science carefully considers and evaluates anomalies in data and evidence.	• Scientific knowledge is based on the assumption that natural laws operate today as they did in the past and they will continue to do so in the future. • Science assumes the universe is a vast single system in which basic laws are consistent.
Science is a Human Endeavor (CCCs)	• Men and women from different social, cultural, and ethnic backgrounds work as scientists and engineers. Scientists	• Scientific knowledge is a result of human endeavor, imagination, and creativity.

NOS Principles	Middle School	High School
	and engineers rely on human qualities such as persistence, precision, reasoning, logic, imagination and creativity. • Scientists and engineers are guided by habits of mind such as intellectual honesty, tolerance of ambiguity, skepticism and openness to new ideas.	• Individuals and teams from many nations and cultures have contributed to science and to advances in engineering. • Scientists' backgrounds, theoretical commitments, and fields of endeavor influence the nature of their findings.
Science Addresses Questions about the Natural and Material World (CCCs)	• Scientific knowledge is constrained by human capacity, technology, and materials. • Science knowledge can describe consequences of actions but is not responsible for society's decisions.	• Science and technology may raise ethical issues for which science, by itself, does not provide answers and solutions. • Science knowledge indicates what can happen in natural systems – not what should happen. The latter involves ethics, values, and human decisions about the use of knowledge. • Many decisions are not made using science alone but rely on social and cultural contexts to resolve issues.

Content Credit: NGSS Lead States, 2013 (www.nextgenscience.org/).

Note: *NGSS Appendix H is a registered trademark of WestEd. Neither WestEd nor the lead states and partners that developed the Next Generation Science Standards were involved in the production of this product, and do not endorse it.

Equity and Antiracist Science Teaching

Frank Niepold reminds us in the Foreword that although science can effectively tell us what and how to teach regarding climate change, we cannot ignore social science data revealing why changes are not taking place. Part of unveiling our underlying beliefs means that we need to take time to sit uncomfortably with how those beliefs influence our actions and decisions with possible bias. There are science teachers out there who believe that English is the superior language, that the American educational system is doing exactly what it was designed to do, and that specific ethnic groups don't have what it takes to pursue STEM careers, etc. Consider carefully how these deficit core beliefs might influence a teacher's everyday decisions. The educational system is made and reaffirmed by people. If we want to disrupt that system in any way, we have to start by looking inwards to see if we are perpetuating the cycle or working to dismantle it.

Ibram X. Kendi (2019) asserts that when it comes to racist or antiracist work, we are all actively engaging in one or the other through our daily decisions. Recall that teachers hold a great deal of power in the classroom when they are able to assign grades, choose which voices to elevate, design curricula, create community, and much more. As we reimagine what science education can be, think about ways that your daily lessons affirm your students' viewpoints, culture, or identity. When you are ready to challenge yourself to critically reflect and work towards equity and antiracist science teaching, consider engaging with the activity in Exhibit 1.7.

Exhibit 1.7 Reimagining Science Teaching as Antiracists

As you read through and think critically about your own practices and beliefs, indicate which questions you find yourself drawn to and wanting to know more about. Consider the following umbrella questions with supportive guiding questions to engage in antiracist science teaching:

- ◆ Where do my students see themselves in this lesson?
  - ✓ Specifically, how are they represented or positioned during today's lesson?
  - ✓ At what point(s) will they co-construct knowledge with me to make sense of what they are learning?

- ✓ Who is driving this lesson (you, your students, both)?
- ✓ Is the lesson culturally relevant or responsive in any way?
- ◆ How does this lesson affirm students' cultural identity to empower them by positioning them as capable contributors?
  - ✓ When you teach this lesson, are students positioned as empty buckets that need to be filled with content knowledge, or as individuals with cultural wealth and knowledge that can contribute in different ways?
  - ✓ Do you know your students' stories or family histories?
  - ✓ How might storytelling be empowering or affirming for students in your class?
  - ✓ Do they see themselves represented in the classroom space?
  - ✓ How important is genuine relationship building in your class? Why?
- ◆ Do I recognize ways in which I have assimilated to or am part of the dominant culture, and how that may potentially be shaping my pedagogical practices and classroom decisions?
  - ✓ First, am I aware of teaching practices that reinforce the dominant culture?
  - ✓ Try thinking of a scientist or engineer. What do they look like? What do they sound like? Do most scientists or engineers look like that person?
  - ✓ Are there benefits to my curriculum being more culturally relevant and responsive for my students?
  - ✓ Is there value in affirming students' cultural identities, their unique ways of thinking, or their different ways of doing science?
  - ✓ When I showcase scientists or engineers, are they diverse and from different cultures around the world?
  - ✓ Do I hold scientists from my country in higher regard than scientists from other countries? Why or why not?

Kendi further defines antiracists as individuals who believe that racial groups are equals that do not need developing (2019). You might be wondering, how might that look like in a classroom setting? Generally, it first begins with teachers that have not taken the time to explore their underlying core values who unintentionally (or intentionally) press upon students the need to assimilate for better life opportunities (succeeding in a job, higher education, etc.). One example is the teaching of "code-switching" to underrepresented minority students (URMs). This is where teachers stress the need to speak or behave in specific ways in the classroom to assimilate. On the surface we mean well when we ask students to correct the way they express their ideas to "sound like scientists," but how might that shape their cultural or STEM identity if they heard that multiple times per day by various school teachers? Why is *what* they are saying discounted because of the *way* they're saying it (even when the ideas or questions are valid)? When they're repeatedly told to change part of themselves, what is the underlying message about who they are or how teachers might see them?

STEM fields *need* more diversity for different perspectives and ideas to advance for the betterment of our planet. Antiracist teachers are those who believe that racial groups are capable contributors with a wealth of knowledge. Code-switching does not affirm students' identity or validate who they are – instead, it is about developing them into who we *believe* they need to be to obtain success in our society (now or later in life). The fact is, many scientists and engineers do not use formal ways of articulating ideas, may or may not speak or even write in English, are diverse from around the world, and communicate scientific studies and inventions in various celebrated languages. We need to think about how these subtle lessons (that have nothing to do with science and everything to do with culture) might be serving as barriers to URMs and STEM.

How empowered would students feel if they knew how cultures around the world have contributed and continue to contribute to STEM? For example, look at the awe-inspiring science and engineering designs that complement nature in Micronesia (an underappreciated Third World country). Back in the year 1200, the Saudeleur Dynasty manipulated inland waterways to transport large basalt stones (some weighing 100,000 pounds) across 25 miles to create the engineering wonder that is Nan Madol (a lost city where the wealthiest and most revered once inhabited as they controlled oceanic trading operations). Advanced science and engineering took place here before the words "science" and "engineering" even came into existence. How we think and feel about our students and their capacities will impact how and what we teach. Are we able to position them as capable contributors while building their capacity and while celebrating their whole selves to achieve what others might believe is impossible?

If we want to reimagine what science teaching can be to support every student, Zaretta Hammond argues that, "We need to be brave enough to interrogate our practice for the sake of our kids (2015)." Just as building relationships with students takes time through small daily interactions, creating an inclusive classroom space that serves as both "windows" into other cultures as well as "mirrors" that reflect students requires intentionality (Sims Bishop, 1990). Being an antiracist is not a destination, but rather a process you engage with as you make everyday classroom decisions ranging from how to deliver a lesson, the lesson design, and the space in which it will all take place.

Putting the Pieces Together

Educational researchers emphasize three main elements needed to support students learning science (National Academies of Science, Engineering, and Medicine, 2018; NRC, 2005). The first element is that teachers need to address students' current understandings about the content (some might refer to this as prior knowledge). If you want to teach students new content, you have to first unveil what they currently know and how they think they know that to unveil their funds of knowledge. If there are initial misunderstandings, consistent opportunities for discourse will let you gauge their depth of knowledge and potential gaps to use as starting points. Remember that if a student has firmly held on to certain misconceptions for long periods of time, it will take more than one class period for them to believe they are wrong just because you say they are. Exhibit 1.8 will provide you with various teacher talk moves to support this element.

Exhibit 1.8 Teacher Talk Moves

Teachers are more likely to support students' ability to connect their current understanding with new concepts by being intentional in their questioning, listening, and affirmation of students' thoughts. This happens effectively through consistent and multiple opportunities for discourse to allow for co-construction of knowledge with students.

Deliberate Questioning

◆ *Encourage students to say more.* This move allows for teachers to gauge how many students feel or think similarly, and allows the student to reflect more deeply about what they are thinking.

◆ *Revoice responses.* By reiterating what the student said, it could bring clarity to the student who hears it back so they can agree or correct their own ideas.

◆ *Ask for reasoning using evidence.* Go beyond a superficial level of conversation to deeper discourse for sense-making by exploring why students think or feel that way about their adopted beliefs.

Listening with Intent

◆ *Powerful paraphrase.* Ask students to restate what their peers are sharing to engage students in conversation and allow for students to affirm each other's ideas.

◆ *Challenge or offer counterexamples.* Are you or students able to push each other by providing counterexamples? Consider also trying this with a student who provides the right answer to unveil their thinking and rationale.

◆ *Wait time.* Provide wait time between questions and responses so students can process information at their own pace. Consider tracking how often you reward the fast thinkers in your class and how to more equitably tend to every learner.

Affirming Students

◆ *Acknowledge.* Understand that your students have these current ideas that stemmed from something very real to them. Start where they are and through questioning, bridge the gap so they can get to where they need to be. Is there something in their response that allows for you to empathize with *why* or how they adopted that idea?

◆ *Set up affirmation opportunities.* Ask a student to share and find students who will affirm the initial student's ideas. This will allow you to see who else shares the same current understanding and why. When you move to asking who disagrees and why, it would be helpful to gently point out that it's okay to be wrong (engaging in the nature of science) because perhaps they didn't have the information or data currently presented and more importantly, they are not alone in how they thought about that old idea and it's understandably why they did.

The second element emphasizes the need for students to understand the nature and process of science. Moving beyond The Scientific Method, this element calls for students to be positioned as scientists and engineers by engaging with different practices to do science. If we are relying on students to use science as a tool to solve real-world complex problems, they need to engage in and understand the true nature and process of science to be scientifically literate. Recall that positioning students as scientists and engineers will greatly depend on your personal beliefs as well. Take time to reflect on your values and beliefs about how students learn in relation to equity and antiracist teaching practices. Exhibit 1.9 will allow for you to tend to equity in your class through challenging reflection.

Exhibit 1.9 Tending to Equity through Personal Beliefs

As you think through how to successfully position students in your class, considering some of the following:

◆ When will students talk with me, each other, or with the whole class in this lesson? Whose voices are being elevated during those moments? Whose ideas carry more value? Whose receiving consistent affirmation in that space?

◆ Who is represented in this curriculum? How are they represented or positioned?

◆ How culturally relevant or responsive is this lesson or curriculum? Are there ways to adapt it to the needs of my students?

◆ Do I hesitate to call on certain students for any reason? Am I willing to explore why I feel that way? Then, am I willing to make any changes (if needed) from what I uncover about my thinking?

◆ Do I find myself constantly asking students to change how they speak or behave? Are those things related to science content, individual behaviors, personal beliefs I have about how students carry themselves, or something else?

◆ When I listen to student responses, am I listening for what is being said/asked, or do I find myself stuck on *how* it's being said/asked? Do I feel that one is more important than the other? Why or why not?

◆ Where might students be able to co-construct knowledge with me in my daily lesson? Is this important to me? Is this important for my students? If I feel that students are not ready for this, how did I determine that? Are these students co-constructing knowledge with other content teachers? Is that important for me to know?

The third element highlights the need for students to participate in metacognition (thinking about their own thinking for sense-making). Allowing for students to reflect and think about their thinking fosters cognitive growth. Much of the NGSS requires for students to provide reasoning or justification for a claim, and allowing for students to question themselves, their decisions, or their peers allows for them to be critical consumers of information. Science relies heavily on credible sources of evidence that allow individuals to accept, reject, or modify claims in light of new information. When students see their teachers or peers modeling the nature and process of science, they begin to truly understand how credible evidence carries more weight than any number of expert opinions. Consider providing opportunities for students to identify their own learning dispositions when you start class. Can you tend to their social and emotional well-being so they have moments to recognize what motivations or challenges are impacting their learning experience today?

Without analyzing your underlying core beliefs and teaching values, you might inadvertently approach the NGSS in ways similar to the old state standards. To ensure that you are successful in taking on climate change for NGSS, consider what you're willing to change and why you would invest time to change it. Just as the NGSS is not a series of checkboxes that may or may not fit into your current curriculum, teaching about climate change to empower students requires more than just temporary strategies or tools. For both, there needs to be a paradigm shift. This chapter is meant to provide you with an overview of what is needed to enhance your science practices to align to NGSS, provide a clear rationale for the paradigm shift, and offer ideas to consider while you reflect on your theories of action.

 Connect with the network of teachers to share your thoughts on education for climate action by leveraging the NGSS. Connect with others starting that are also starting their learning journey at www.empoweredscienceteachers.com (Book Resources → Chapter 1 → Discussion Board)

Collective Voices for Climate Change Education

Kristen Iverson Poppleton (Senior Director of Programs at Climate Generation)

Climate change education should be centered in equity and justice. In the science classroom this means teaching in the context of the whole child – understanding and listening to your students' lived experiences and making everyday life connections. It means making the subject of climate change pervasive, not just in one grade level, one lesson or one unit. This also means looking for opportunities to connect with other subject areas, especially civics and language arts. Finally, climate change education needs to be at least 50 percent focused on solutions and opportunities for students to take action, innovate solutions, and take leadership.

Loyda Ramos (Education Outreach Manager at TreePeople)

If teachers can help students identify their specific passion within the larger context of climate change (such as water conservation, source reduction, environmental justice, etc.), then share guidelines on how to advocate on local levels in their communities and with their local government officials – it will help students develop their individual voice about something they feel strongly about and then learn how to work within the system to affect change.

Additional Teacher Resources

Access "How Students Learn (2005)" –
 Bit.ly/HSLSCIENCE

Get teaching tools with UC Berkeley's Resources –
  Bit.ly/CALTOOL2
Learn about the Nature of Science in the NGSS –
  Bit.ly/NGSSNOS
Learn about the Process of Science –
  Bit.ly/Process OfScience
See your school's CAST results –
  Bit.ly/CASTschool
Watch Haas' (2020) Science in the Virtual Pub segment –
  Bit.ly/HaasSVP

References

Kendi, I. X. (2019). *How to be an antiracist.* First Edition. New York: One World.

Hammond, Z. (2015). *Culturally responsive teaching and the brain: Promoting authentic engagement and rigor among culturally and linguistically diverse students.* Thousand Oaks, CA: Corwin.

Hass, D. (2020). Reinventing the educational system in a time of disruption: Kick-off for science in the virtual pub [Video]. YouTube. www.youtube.com/watch?v=HQOaGLgoXGk.

Miller, J. D. (2016). Civic scientific literacy in the United States in 2016. https://science.nasa.gov/science-red/s3fs-public/atoms/files/NASA%20CSL%20in%202016%20Report_0_0.pdf.

National Academies of Sciences, Engineering, and Medicine (2018). *How people learn II: Learners, contexts, and cultures.* Washington, DC: The National Academies Press. https://doi.org/10.17226/24783.

National Research Council (2005). *How students learn: Science in the classroom.* Washington, DC: The National Academies Press. https://doi.org/10.17226/11102.

National Research Council (2013). *Next Generation Science Standards: For states, by states.* Washington, DC: The National Academies Press. doi: 10.17226/18290.

National Science Teachers Association (NSTA) 2016. *NSTA Position Statement: The National Science Teachers Association.* https://static.nsta.org/pdfs/PositionStatement_NGSS.pdf.

NGSS Lead States (2013). *Next Generation Science Standards: For states, by states.* Washington, DC: The National Academies Press.

Sims Bishop, R. (1990). Mirrors, windows, and sliding glass doors. *Perspectives: Choosing and Using Books for the Classroom*, 6(3), ix–xi.

2

Leveraging the NGSS
for Climate Change

Read this when:

- ♦ *You're ready to learn and engage in more NGSS-aligned instruction.*
- ♦ *You're considering the role of climate change in the NGSS and wondering about how the two complement each other.*
- ♦ *You're ready to take action on what you have learned to see results in your classroom.*

A Story about Root-Bound Plants

Ron Finley is known as "The Gangsta Gardener" from South Los Angeles, who works to provide accessible healthy food options to inner city neighborhoods through community work, advocacy, and education. Finley has an online course to greenify spaces and one lesson focuses on helping people rescue "root-bound plants." Root-bound plants are ones that grow inside small containers that you can typically find at the store. You can identify a root-bound plant by looking for roots that are growing towards the top layer or growing out of the bottom holes of the pot. As a result, we know that these plants have spent a significant amount of time in their small container. When the roots begin to fill the shape of the container, it causes these plants to be smaller and less healthy compared to plants that have more space. Interestingly, the container can determine how much and how well the plants are able to grow.

When you think about Finley's lesson on root-bound plants, consider how it parallels with students and the environment in which they learn. Given that there are many uncontrollable external factors and challenges that also play a role in their educational trajectory, there is no denying the powerful difference that every teacher makes. Recall how much power teachers have in the classroom. What students learn and develop in your science class is also determined by what you are ready to take on or what you decide to push off. Based on your personal beliefs, you have the power to determine if students are capable of driving the lesson, will engage in meaningful discourse (not to regurgitate information), share their cultural wealth, do STEM, etc. This chapter will continue to deconstruct the NGSS to leverage the framework for climate science to support your successful classroom implementation of the material.

A Second Look at the NGSS for Climate Science

When thinking about the shift to NGSS, we need to accurately identify the problem in order to come up with the right solution. When you listen carefully to how people frame the issues regarding NGSS, it reveals a great deal about their values and beliefs that influence their actions. Without a paradigm shift, teachers might inadvertently approach the NGSS as disconnected components they force into current lessons. Although it is the easiest approach, we know this type of approach will not get teachers the results they are hoping for. More importantly, this type of approach does not take into consideration the pedagogical shifts needed to implement NGSS curricula not spelled out in a lesson plan. Anyone can pull "NGSS" resources off the internet, but how they enact the lesson reveals another level of alignment needed for the new framework. It will be hard work to revamp your curriculum (especially if you're thinking of using climate change as the driving force to engage and empower students), but you will also find it more meaningful as your students develop strong scientific identities, skills, and desire to be the change agents we need. Consider participating in Exhibit 2.1 to see if your beliefs about NGSS have shifted.

Exhibit 2.1 What's the Problem?

The majority of the nation has shifted from the old science framework to the NGSS. What are some reasons for why you think that might be (select your *top three answers*)?

◆ The standards change every decade or so. It was just a matter of time.

◆ There were too many content requirements in the old framework and we needed more depth.

◆ Students were technically not required to *do* science with the old framework.

◆ Equity was not a core principle in the old framework (accessible science for all).

◆ As a nation, students were not as prepared as they could be to go into STEM fields.

◆ The old framework typically benefitted students who did well with rote memorization and recall of concepts.

◆ We have more research revealing new information on how students learn science now (including culturally relevant and responsive pedagogy, considering the role of culture and learning, social and emotional well-being of students, need for equity, etc.).

◆ There were resource inequities between schools accountable for the same standards.

◆ Under the old framework, there was little to no coherence between science classes at different levels (across teachers at the different grade/content levels).

◆ Under the old framework, there was little to no coherence between science classes at the same levels (between teachers at the same grade/content level).

- ◆ Textbooks used to supplement information are archaic and we needed new ones.
- ◆ Students were positioned as empty buckets or sponges that we should fill with information.
- ◆ Under the old framework, lessons rarely connected to something larger, are meaningful to students personally, or the natural world.
- ◆ Students needed to start developing scientific literacy skills.
- ◆ Students needed more opportunities for sense-making rather than memorization of facts.
- ◆ The Scientific Method does not reflect the true process or nature of science.
- ◆ Other: _____

Depending on the boxes you checked, that reveals the lens that you might be approaching the problem from. For example, if I *only* checked the first box, I might be thinking that the NGSS is just another state requirement. I might not also value the components of the NGSS because I view it as another requirement (a change just to change something rather than having a meaningful purpose). I also might be thinking that "This too shall pass," so if I wait it out long enough new standards will replace this one so there's no point in revamping my curriculum. Lastly, I might not see any flaws in the old framework because the majority my students did well in "mastering" the content. You can see how these beliefs will shape the actions that are taken by the teacher (a.k.a. their TOA) in shifting towards the NGSS.

 Think about your values and beliefs about teaching and learning. What actions might you take depending on the beliefs you have about shifting to the NGSS?

 Think about your science colleagues. What actions do you think they are taking regarding the NGSS, which might reveal some of their beliefs about the

new framework? Thinking about this will help you to better understand how others might operate and lead to better collaboration efforts in the future.

In this book, the top three reasons identified for the shift to NGSS includes (but is not limited to) the following:

1. Treating science as a creative and innovative process to solve everyday problems, rather than only rediscovering what we already know to be true.
2. Using climate change as the vehicle to learn about the nature of science to further develop students' scientific literacy skills.
3. Positioning all students as knowledgeable and capable doers of science that can contribute to STEM fields.

Solutions provided by this book to address the above include education on the process and nature of science to shift the way we think about science teaching through NGSS for climate change (Chapters 1 and 2), beginning support on integrating climate science into your curriculum for scientific literacy skills (Chapters 3 and 4), and providing advanced resources to empower your students to use science to address issues in their communities in which they have the answers for (Chapters 5 and 6). To also have a better understanding between the old and new science frameworks, explore Table 2.1 for an overall comparison.

A Fresh Take on the NGSS

The new science framework calls for teachers to integrate science content and solutions to combat societal problems (the perfect catalyst needed for education leading to climate action). There are three major components of the NGSS that include: the science and engineering practices (SEPs), disciplinary core ideas (DCIs), and crosscutting concepts (CCCs). The SEPs refer to what scientists do

TABLE 2.1 Comparing the Frameworks

The Old Framework	The New Framework (NGSS)
1. Held students accountable for a wide range of content. 2. Rewards students skilled at rote memorization. 3. Thinks of student learners as sponges that mainly absorb content information. 4. Excluded engineering. 5. Excluded climate science. 6. Stresses The Scientific Method. 7. Often assessed through multiple-choice emphasizing content recall. 8. Students are not accountable for applying content knowledge to the real world. 9. There's no emphasis on cross-curricular or interdisciplinary teaching. 10. There is little accountability and unclear expectations for what students will do with the content. 11. Does not consider research on how students learn new science content. 12. Lacks culturally relevant approaches to teaching science. 13. Emphasis is on knowing about science.	1. Holds students accountable to performance expectations that calls for three-dimensional learning. 2. Positions students as capable critical thinkers, scientists, engineers, and agents of change. 3. Thinks of students as knowledgeable contributors and values their cultural backgrounds and communities. 4. Integrates engineering components. 5. Integrates climate science. 6. Introduces the true process of science. 7. Emphasizes scientific literacy through learning about the nature of science. 8. Assesses students' critical thinking skills to apply content rather. 9. Recognizes the complexity of science and calls for an integration of mathematics and English language arts. 10. Aims to prepare all students for college, career, and citizenship. 11. Takes into account how students learn through cyclical and iterative processes (example: sense-making through phenomenon-based instruction). 12. Views culturally relevant pedagogy as essential to engaging and empowering students. 13. Emphasis is on doing science.

Note: The NGSS is a registered trademark of WestEd. Neither WestEd nor the lead states and partners that developed the Next Generation Science Standards were involved in the production of this product, and do not endorse it.

as they investigate phenomena around them and how engineers design or create systems as a response. The DCIs are key content ideas that build off one another through and across grade bands. Lastly, the CCCs are the connections between major science domains (such as Earth and Space Science, Physical Science, Life Science, and Engineering). When seamlessly intertwined,

TABLE 2.2 NGSS Three-Dimensional Learning

Science & Engineering Practices	Disciplinary Core Ideas	Cross Cutting Concepts
• Ask questions and define problems • Plan and carry out investigations • Use mathematics and computational thinking • Engage in an argument from evidence • Develop and use models • Analyze and interpret data • Construct explanations and design solutions • Obtain, evaluate and communicate information.	• Life Sciences • Physical Sciences • Earth & Space Sciences • Engineering, Technology, and Application of Science	• Patterns • Scale, proportion, and quantity • Energy and matter • Cause and effect • Systems and system models • Structure and function • Stability and change

Content Credit: NGSS Lead States, 2013 (www.nextgenscience.org/)

CCC → 3D learning

the three components provide students with three-dimensional learning (refer to Table 2.2). This approach supports building students' capacity as well-informed decision makers on STEM policy by increasing their scientific literacy skills.

Nearly one-third of the NGSS relates directly or indirectly to climate change content, and there are many opportunities to incorporate climate science coherently within and across grade levels. Since the majority of teachers don't receive formal training on teaching climate science for NGSS, it's important to highlight climate science as it relates to the new framework. Although not explicit in elementary standards, primary teachers do teach about the cycling of matter and Earth's spheres as a basis of climate science among other concepts. For secondary standards, climate science is written through different Performance Expectations (PE) and DCI (e.g. including Weather and Climate, Human Impact, Human Sustainability, Earth's Systems, Earth & Human Activity, etc.), energy is a CCC woven throughout all courses, and pushing students to develop solutions for climate related issues they will face requires the SEPs and Engineering, Technology, and Application of Science standards (ETS). Please refer to *Additional Teacher Resources* at the end of the chapter to access several resources that outline climate science in the NGSS.

Along with understanding where climate science is embedded in the NGSS, it's also important to note that young people across the country want support in learning about and leading change for the climate crisis. According to the US National Action for Climate Empowerment Strategic Planning Framework, public empowerment is necessary to meet the on-going challenges of climate change (Bowman & Morrison, 2020).

> The solutions to the negative effects of climate change are also the paths to a safer, healthier, cleaner and more prosperous future for all. However, for such a future to become reality, citizens of all countries, at all levels of government, society and enterprise, need to understand and be involved.
>
> (Paas & Goodman, 2016)

Pressing issues such as the climate crisis are predicted to lead in education as environmental justice issues become the center of attention this century. With this in mind, climate change serves as the ideal topic because it directly impacts people and will require innovative, diverse, and creative ways of thinking to mitigate.

The Role of Climate Literacy and NGSS

Although climate science is complex and multifaceted, there are major scientific components that every science teacher should address with students that have been identified by NGSS and field experts. The NGSS Earth and Space Science PEs push students to understand system interactions that influence weather and climate, analyze and interpret geoscience data that drives climate change, unpack the significant interdependencies between humans and Earth's systems by looking at the impacts of natural hazards, critically examine our dependency on natural resources, and recognize the impact human activities have on the environment. Students demonstrating content mastery should be able to develop and use models, analyze and interpret data, apply mathematics and computational thinking, construct explanations, design solutions, and engage in argumentation using evidence. The development of this section in the NGSS was strongly influenced by several literacy principles including the Climate Literacy Principles, which outlines major learning objectives for students.

Similar to the NGSS, *The Teacher-Friendly Guide to Climate Change* (TFG) assembled by Zabel, Duggan-Haas, and Ross (2017), also identified five major concepts that all students should understand about climate change. The authors interviewed expert climate scientists, social scientists, science educators, and climate journalists who agreed on big ideas along with two overarching questions educators should consider when teaching climate change (Zabel, Duggan-Haas, & Ross, 2017). Just like the NGSS, they also acknowledge the importance of starting with the Climate Literacy Principles to focus on what all students should understand. The first big idea is to acknowledge that climate change is a real and very serious problem that our society faces now and in the coming centuries. Second, climate change is caused by anthropogenic factors, especially when it comes to energy use. Anthropogenic is a term used to describe the impact on Earth as caused or influenced by human activities. Third, it is important to understand that humans can take actions to mitigate climate change and its impacts. Fourth, there is a need for mathematical thinking to understand time, scale, models, and maps in depth related to climate change. Lastly, experts argue that understanding that Earth is a system of complex systems is the most important concept because many subjects are connected in explaining the climate crisis (note that the four big ideas mentioned previously rely on this one). The two overarching questions draw upon the Nature of Science (NOS) to encourage students to examine how scientists know what they know to understand the scientific process, and how that information informs decision-making. Table 2.3 outlines the Climate Literacy Principles, climate change content in the NGSS, and the recommendations put forth by the Paleontological Research Institution to show coherence.

TABLE 2.3 Connections for Climate Change Curriculum Design

Climate Literacy Principles (NOAA, 2009)	Next Generation Science Standards (NGSS, 2013)	Paleontological Research Institution (TFG, 2017)
Essential Principle 1: The Sun is the Primary Source of Energy for Earth's Climate System.	Core Idea ESS1: Earth's Place in the Universe • ESS1.C: The History of Planet Earth	Big Idea 1: Climate Change is a real and serious problem facing global society in the coming decades and centuries.
Essential Principle 2: Climate is regulated by complex interactions among components of the Earth system.	Core Idea ESS2: Earth's Systems • ESS2.A: Earth Materials and Systems • ESS2.C: The Roles of Water in Earth's Surface Processes • ESS2.D: Weather and Climate • ESS2.E: Biogeology	Big Idea 2: Climate change in recent decades is primarily caused by human activities, especially as related to energy use.
Essential Principle 3: Life on Earth depends on, is shaped by, and affects climate.	Core Idea ESS3: Earth and Human Activity • ESS3.A: Natural Resources • ESS3.B: Natural Hazards • ESS3.C: Human Impacts on Earth Systems • ESS3.D: Global Climate Change	Big Idea 3: Humans can take actions to reduce climate change and its impacts.
Essential Principle 4: Climate varies over space and time through both natural and man-made processes.	Core Idea ETS1: Engineering Design • ETS1.A: Defining and Delimiting an Engineering Problem • ETS1.B: Developing Possible Solutions • ETS1.C: Optimizing the Design Solution	Big Idea 4: To understand (deep) time and the scale of space, models and maps are necessary.
Essential Principle 5: Our understanding of the climate system is improved through observations, theoretical studies, and modeling.	Core Idea ETS2: Links Among Engineering, Technology, Science, and Society • ETS2.A: Interdependence of Science, Engineering, and Technology • ETS2.B: Influence of Engineering, Technology, and Science on Society and the Natural World	Big Idea 5: The Earth is a system of complex systems.

Climate Literacy Principles (NOAA, 2009)	Next Generation Science Standards (NGSS, 2013)	Paleontological Research Institution (TFG, 2017)
Essential Principle 6: Human activities are impacting the climate system.		Overarching Question 1: How do we know what we know?
Essential Principle 7: Climate change will have consequences for the Earth system and human lives.		Overarching Question 2: How does what we know inform our decision-making?

Note: NOAA is part of the U.S. Department of Commerce and their mission is to keep citizens informed about the changing environment using big data collection systems to monitor Earth's systems. The NGSS has several Earth and Space Science (ESS) standards as well as Engineering, Technology, and Application of Science (ETS) standards for educators to cover. Lastly, *The Teacher-Friendly Guide to Climate Change* (TFG) synthesizes major ideas for educators to consider as they develop curricula around climate change.

Climate Change as a Socioscientific Issue

Climate change serves as the ideal vehicle to learn about the NGSS, NOS, and scientific literacy. Looking at ethical issues related to the climate crisis, it's important to remember that climate change impacts people differently. Students need to learn how humans are the drivers for anthropogenic climate change, how they are directly impacted, and how Earth's systems are changing. We are already experiencing the impacts of climate change and we can support students as agents of change by providing the tools and skills needed to explain and take action on this devastating phenomenon. So how can we teach about climate change through the NGSS to develop scientifically literate students? Prior research tells us that teaching climate change as a socioscientific issue is key (Hestness, McGinnis, Riedinger, & Marbach-Ad, 2011; Hestness et al., 2014; Holthuis, Lotan, Saltzman, Mastrandrea, & Wild, 2014; Matkins & Bell, 2007; Sadler, Chambers, & Zeidler, 2004).

Socioscientific issues (SSI) are complicated, open-ended, and controversial without straightforward solutions (Sadler et al., 2004). As an SSI, climate change is a politically controversial issue that students and the public often hear about in the media. When the majority of teachers shy away from teaching about the climate crisis for various reasons, the media and Internet become primary sources of information for climate change (Caranto & Pitpitunge, 2015; Carter & Wiles, 2014; Hestness et al., 2014; Kolstø, 2001). It is important to note that students will come across climate change information whether or not teachers include or omit the content in the curriculum (Kolstø, 2001; Sadler et al., 2004). By being deliberate with *how* we teach about the climate crisis, we can help students develop as 21st Century scientifically literate citizens capable of pushing back against disinformation campaigns or misleading information put forth by the media (Bunten & Dawson, 2014; Lambert & Bleicher, 2013).

Although the scientific data supporting climate change is not widely disputed, most educators and the public still doubt scientific consensus around the issue (Cook et al., 2013; Leiserowitz, Maibach, Roser-Renouf, Rosenthal, & Cutler, 2017; Somerville & Hassol, 2011). Rather than teaching the data that supports climate science, some teachers avoid or teach their personal opinion on the subject (Hestness et al., 2014; McCaffrey, 2015; Plutzer et al., 2016). This is problematic since the data supporting climate science is not disputed, but is so politically controversial that people think it is debatable. Socioscientific issues include disagreements related to conflicting evaluations of the validity or trustworthiness of science-related claims. Teaching climate change as an SSI allows for students to build their analytical skills through questioning, data analysis and interpretation, discourse, and more.

The SSI framework encourages teachers to address social and ethical dimensions of climate change to support students as agents of change. We are seeing an uprising across the world with young people leading protests in their cities demanding that governments take action to bend the curve. The environmental issues that students feel empowered to act upon and take more interest in are related to climate change's controversial nature and their right to learn the science of climate change. An example of this is when students learn about disparate racial, social, and environmental injustices (directly resulting from the climate crisis) through community data analysis by zip code. Seeing the direct impact, students may also become highly interested to better understand how they can influence local policy and ways to involve their community in taking action. They will want change that they can see and feel for their communities, and teachers can provide safe spaces and opportunities for students to hold these conversations around complex issues.

The consequence of not addressing the political interests, social values, or ethical implications underlying the climate crisis is that we *disempower* students (Hodson, 2003; Plutzer et al., 2016). Hodson (2003) argues that avoiding judgments in science is not possible since values are embedded in the science curriculum whether teachers recognize it or not. It is also important to note that values can also be promoted by content that is omitted from the classroom just as much as what is included. Hodson (2003) adds that the purpose of education should be to empower individuals to critically analyze society and values needed to sustain it. Students need to ask what can be changed and how they can make those changes to address direct community impacts and ensure environmental sustainability. I invite you to complete the activity in Exhibit 2.2 to gauge your level of comfort and willingness to address ethical dimensions related to the climate crisis.

Exhibit 2.2 Where Am I Now and Where Do I Want to Be?

Research reveals that teaching climate change as an SSI will empower your students to be agents of change in their own communities. It is important to gauge where you are in regards to addressing the ethical dimensions that relate to climate change to build student capacity.

Consider answering the following questions when you're ready:

1. How often do you currently teach about social, racial, or political science issues in your class now? Why?
2. Is there a possibility for interdisciplinary work with other subject matters on climate change (Social Science, Mathematics, or Language Arts subjects)? Is that a priority for you?

3. How often do you include current science events directly related to climate change in your class? Why?

4. How important is it to provide students with discourse opportunities (such as argumentation) on relevant science-issues? Why?

5. Can you imagine integrating opportunities for students to address ethical issues around climate change in the future (Some example issues include access to clean water, pollution, human impacts, green technology, consumerism driving the economy, climate change policy, or nuclear power)?

6. How comfortable are you with teaching content that you might not have all the answers to? Does that determine whether or not you include it in your curriculum?

7. Climate change is tentative in nature because there are no clear solutions, and the data is still being collected showing the variety of impacts on Earth. How comfortable are you teaching about this content?

8. Scholars often argue that the educational system is about reinforcing the status quo (i.e. U.S. education was never meant for marginalized underrepresented minorities and only recently has been challenged). How much time and energy are you willing to invest to disrupt science education to provide accessible lessons to all students?

9. What do you think is the purpose of science education? Is your current instruction aligned to your vision?

Reflect on the answers you wrote above, and consider where you would like to shift your practices. Remember that students will be confronted by the ethical dimensions of climate change whether or not you address them in your class. How important is the climate crisis to you as it informs your curriculum design decisions? How will you help them use science to make informed decisions regarding those claims? How might students do science and engineering to mobilize on this issue?

 Connect with the network to see what other teachers are doing to magnify their impact on education for climate action. See how others are addressing the ethical dimensions of the climate crisis at www. empoweredscienceteachers.com (Book Resources → Chapter 2 → Discussion Board)

Deeper Dive into SSI

The SSI framework consists of three core components while also acknowledging the role of *classroom environments* (any norms and expectations established) and *peripheral influences* (external factors that might influence the outcome). The three core aspects presented by Presley et al. (2013) are *Design Elements, Learner Experiences,* and *Teacher Attributes.* The researchers argue that the SSI-based framework supports scientific literacy by taking into account "real-life" scientific situations that may be influenced by other factors such as politics, social, or ethical issues. This perspective is also consistent with the Science and Engineering Practices of the NGSS.

Presley et al. (2013) emphasize that each of the core aspects of the SSI framework identify important features of teaching and learning. *Design Element* refers to creating instruction around a compelling issue, presenting it first, scaffolding higher-order practices, and providing a culminating experience for students. The central issue must be a compelling social issue with clear connections to science (such as climate change or evolution). The instruction is based on providing real-world contexts so that students will gain a deeper understanding of the science while developing skills needed to make decisions beyond the classroom. *Learner Experiences* focuses on further developing students' higher-order practices (reasoning, argumentation, decision-making, or positionality), confronting scientific theories, collecting and analyzing data, and considering other dimensions that may have an influence (social, political, or economical). This core aspect also pushes students to confront ethical dimensions of SSI while applying the NOS principles in their analysis. *Teacher Attributes* looks at how familiar the teacher is with the SSI, sees teachers as learners, and considers their willingness to deal with uncertainties that may arise from addressing these open-ended controversial topics. Presley et al. (2013) stress that successful SSI instruction relies heavily on teacher awareness of any social considerations related to the topic (e.g. to teach climate change effectively you have to consider the politically controversial nature of the topic that influences students' current understandings).

Connection to the Nature of Science

Integrating controversial SSI such as climate change into your science curriculum is an opportunity for you to teach NOS principles and NGSS. Recall that the NOS calls for students to examine a topic's empirical evidence, social and cultural embeddedness, and tentative nature (Sadler et al., 2004). These major components of the SSI framework push students to utilize major NOS themes to formulate ideas regarding climate change on evidence. Researchers argue that students' are likely to accept or take action on controversial SSI after understanding the nature of scientific knowledge (Carter & Wiles, 2014; Khishfe & Lederman, 2006; Kolstø, 2001; Sadler, Barab, & Scott, 2007). Although this approach can effectively strengthen scientific literacy, research reveals that teachers need more learning opportunities that effectively integrate the NOS through SSI to help students become informed decision makers (Kolstø, 2001; Lee et al., 2012; Matkins & Bell, 2007; Sadler et al., 2004, 2007).

As previously addressed, the old framework does not address the NOS, which is necessary for students to successfully confront politically controversial topics they will encounter in their daily lives (Caranto & Pitpitunge, 2015; Hodson, 2003; Lambert & Bleicher, 2013; Shea, Mouza, & Drewes, 2016). Teaching students science through SSI is necessary to help students align their thinking with the scientific community and to become informed and engaged citizens (Sadler et al., 2007). Individuals who take action on SSI have a deep and personal understanding of the issue and feel they have a personal investment in addressing or solving the issue (Hodson, 2003; Kolstø, 2001). When students become scientifically literate by understanding the NOS, they will be equipped with the skills required to successfully mobilize for environmental and climate justice.

Moving Forward with Confidence

Looking at the available resources, it is important to acknowledge that there is no curriculum that will work for every science teacher. Science curricula should be tailored to the needs of students, their community, and responsive to their lived realities or experiences. It also needs to be respectful of teachers as pedagogical leaders that know their schools and students, give the freedom to approach teaching through different phenomena, and allow for multiple ways to gauge students' progress. If we are going to empower students to take action on the climate crisis, we need to teach the science of climate change and provide opportunities to engage in argumentation from evidence about the social and ethical implications. When you are ready for implementation support, the next chapters serve as learning, planning, and teaching guides to help you integrate climate science through NGSS.

Track your professional growth by re-examining your beliefs about teaching and learning using the three dimensions of the SSI Framework.

Tracking Your Professional Growth

Student learning experiences in my class	To what extent do I agree or disagree with this statement. What are my current beliefs or values regarding the statement? Why?
Students often discuss policies related to science.	
Students have plenty of opportunities to collect and analyze scientific data or information.	
Students often discuss ethical issues related to science.	
Students engage with the Nature of Science principles.	
Students learn about and engage with the true Process of Science.	
Students co-construct knowledge with me every day.	
Students drive the instruction in my class as capable contributors and do-ers.	

Curriculum design elements I currently value	To what extent do I agree or disagree with this statement. What are my current beliefs or values regarding the statement? Why?
I currently build all lessons/units around anchoring or investigating phenomena.	
I present climate change issues at the start of each unit or lesson.	
Students often engage in argumentation and making claims based on evidence.	
Students engage meaningful discourse opportunities every day.	
My lessons/units are centered around real-world issues that are directly related to my students' lives or community.	
Students often use media/ technology to connect classroom content to the natural world.	

My current teacher attributes	To what extent do you agree or disagree with this statement. What are your current beliefs or values regarding the statement? Why?
I have to know everything about a particular issue before teaching about it.	
I know everything about climate science.	
I feel comfortable admitting to students when I don't know the answer to their question.	
I am comfortable teaching about open-ended issues where I cannot predict student responses.	
I have to feel like the expert in the room.	
I often experience imposter syndrome even for topics that I have strong expertise in.	
I have a strong understanding of the Nature of Science principles.	
I have a strong understanding of the NGSS framework.	

Refer back to the **Book Introduction** where you initially tracked your professional growth. Reflect on how your beliefs might have changed and why that might be. How might your responses unveil more about your teaching disposition?

Collective Voices for Climate Change Education

Lin Andrews (Director of Teacher Support at the National Center for Science Education)

As NCSE began revamping lesson plans with our teacher ambassadors, we wanted our first lesson to make our strongest statement and be the one lesson that all teachers incorporate into their curriculum. The new title sums up what every teacher needs to be telling their students: "It's Real, It's Us, It's Bad, There's Hope." Of these four statements, teachers must begin working with students to understand there is **hope.** *We can solve this crisis, we can change the path we are currently traveling, but it will take the next generation of scientifically literate citizens to do it. Our students are our greatest hope.*

Shelly Backlar (VP of Programs at Friends of Los Angeles River)

When looking at climate change, it can be daunting when we look at the big picture. How can we make a difference on a global scale? In our program, we focus on students' local environment – both natural and built – with an emphasis on how human impact shapes our landscape. While we focus on the Los Angeles River, using a watershed approach makes the concepts we teach relevant to other regions. We let students know that their **actions** *now and in the future make a difference. While we strive for a rewilded River that benefits both humans and habitat, we let students know how opportunities to recreate wetland habitat not only have a positive impact us and wildlife, wetland habitat addresses many big picture issues such as: water quality, groundwater recharge, water supply, flood protection, even carbon sequestration and fire prevention. Fostering science education and stewardship makes for informed, eco-literate citizens who comprise a constituency for change.*

Additional Teacher Resources

Access the NGSS standards by topic –
   Bit.ly/NGSSTOPIC
Download The Teacher Friendly Guide to Climate Change –
   Bit.ly/PRITFG
Learn about achieving climate and environmental stability –
   Bit.ly/BigCCReport
Learn about climate change in the NGSS –
   Bit.ly/NGSSGCC
Learn about Ron Finley and his work –
   Bit.ly/RonFinleyRB
MADE CLEAR's outline of climate science in NGSS –
   Bit.ly/MADECLEARGCC
Read about NOAA's Climate Change Principles –
   Bit.ly/CCNOAA
Read about The Socioscientific Issues Framework –
   Bit.ly/SSIFRAME
Review the 2020 US ACE Strategic Planning Framework –
   www.aceframework.us/

References

Bowman, T., & Morrison, D. (2020) An ACE national strategic planning framework for the United States [Online]. Created in collaborative reflection with the U.S. ACE Community. Available at http://aceframework.us.

Bunten, R., & Dawson, V. (2014). Teaching climate change science in senior secondary school: Issues, barriers and opportunities. *Teaching Science*, 60(1), 10.

Caranto, B. F., & Pitpitunge, A. D. (2015). Students' knowledge on climate change: Implications on interdisciplinary learning. In *Biology Education and Research in a Changing Planet* (pp. 21–30). Singapore: Springer.

Carter, B. E., & Wiles, J. R. (2014). Scientific consensus and social controversy: Exploring relationships between students' conceptions of the nature of science, biological evolution, and global climate change. *Evolution: Education and Outreach*, 7(1), 6.

Cook, J., Nuccitelli, D., Green, S. A., Richardson, M., Winkler, B., Painting, R., ... & Skuce, A. (2013). Quantifying the consensus on anthropogenic global warming in the scientific literature. *Environmental Research Letters*, 8(2), 024024.

Hestness, E., McDonald, R. C., Breslyn, W., McGinnis, J. R., & Mouza, C. (2014). Science teacher professional development in climate change education informed by the next generation science standards. *Journal of Geoscience Education,* 62(3), 319–329.

Hestness, E., Randy McGinnis, J., Riedinger, K., & Marbach-Ad, G. (2011). A study of teacher candidates' experiences investigating global climate change within an elementary science methods course. *Journal of Science Teacher Education*, 22(4), 351–369.

Hodson, D. (2003). Time for action: Science education for an alternative future. *International Journal of Science Education*, 25(6), 645–670.

Holthuis, N., Lotan, R., Saltzman, J., Mastrandrea, M., & Wild, A. (2014). Supporting and understanding students' epistemological discourse about climate change. *Journal of Geoscience Education*, 62(3), 374–387.

Khishfe, R., & Lederman, N. (2006). Teaching nature of science within a controversial topic: Integrated versus nonintegrated. *Journal of Research in Science Teaching*, 43(4), 395–418.

Kolstø, S. D. (2001). Scientific literacy for citizenship: Tools for dealing with the science dimension of controversial socioscientific issues. *Science Education*, 85(3), 291–310.

Lambert, J. L., & Bleicher, R. E. (2013). Climate change in the pre-service teacher's mind. *Journal of Science Teacher Education*, 24(6), 999–1022.

Lee, H., Chang, H., Choi, K., Kim, S. W., & Zeidler, D. L. (2012). Developing character and values for global citizens: Analysis of pre-service science teachers' moral reasoning on socioscientific issues. *International Journal of Science Education*, 34(6), 925–953.

Leiserowitz, A., Maibach, E., Roser-Renouf, C., Rosenthal, S., & Cutler, M. (2017). *Climate change in the American mind: May 2017*. Yale University and George Mason University. New Haven, CT: Yale Program on Climate Change Communication.

Matkins, J. J., & Bell, R. L. (2007). Awakening the scientist inside: Global climate change and the nature of science in an elementary science methods course. *Journal of Science Teacher Education*, 18(2), 137–163.

McCaffrey, M. S. (2015). *Climate smart & energy wise: advancing science literacy, knowledge, and know-how*. Thousand Oaks, CA: Corwin.

NGSS Lead States. 2013. *Next Generation Science Standards: For states, by states*. Washington, DC: The National Academies Press.

Paas, L., and Goodman, D. (2016). *Action for climate empowerment: Guidelines for accelerating solutions through education, training and public awareness*. Paris and Bonn: United Nations Educational, Scientific and Cultural Organization and the Secretariat of the United Nations Convention on Climate Change, p. 2. Available at https://unfccc.int/sites/default/files/action_for_climate_empowerment_guidelines.pdf

Plutzer, E., McCaffrey, M., Hannah, A. L., Rosenau, J., Berbeco, M., & Reid, A. H. (2016). Climate confusion among US teachers. *Science*, 351(6274), 664–665.

Presley, M. L., Sickel, A. J., Muslu, N., Merle-Johnson, D., Witzig, S. B., Izci, K., & Sadler, T. D. (2013). A framework for socio-scientific issues based education. *Science Educator*, 22(1), 26.

Sadler, T. D., Barab, S. A., & Scott, B. (2007). What do students gain by engaging in socioscientific inquiry? *Research in Science Education*, 37(4), 371–391.

Sadler, T. D., Chambers, F. W., & Zeidler, D. L. (2004). Student conceptualizations of the nature of science in response to a socioscientific issue. *International Journal of Science Education*, 26(4), 387–409.

Shea, N. A., Mouza, C., & Drewes, A. (2016). Climate change professional development: Design, implementation, and initial outcomes on teacher learning, practice, and student beliefs. *Journal of Science Teacher Education*, 27(3), 235–258.

Somerville, R. C., & Hassol, S. J. (2011). The science of climate change. *Phys. Today*, 64(10), 48.

The Essential Principles of Climate Literacy. (2009). NOAA Climate.gov. Retrieved October 15, 2020, from www.climate.gov/teaching/essential-principles-climate-literacy/essential-principles-climate-literacy

Zabel, I. H. H., D. Duggan-Haas, & R. M. Ross, eds. (2017). *The teacher-friendly guide to climate change*. Ithaca, New York: Paleontological Research Institution.

Part 2

Developing Scientific Literacy Using Climate Science

3

Climate Change Is Complex, Where Do I Start?

Read this when:

- ◆ *You're ready to learn the fundamentals of climate science.*
- ◆ *You need guidance on what to teach regarding climate change.*
- ◆ *You want to know research-based approaches to successfully teach about climate science.*

The Environmental and Climate Change Literacy Project Summit (ECCLPS)

In December 2019, key stakeholders convened to discuss the future of K-16 education at UCLA, and the importance of integrating environmental and climate literacy. Stemming from Dr. Ram Ramanathan's extensive work on climate change, 220 stakeholders (senators, mayors, UC/CSU presidents, deans of colleges, program directors, and educational leaders) gathered to discuss ways to support climate change education efforts for California. The goal set by the steering committee was to educate 500,000 high school students per year to become literate in environmental and climate change issues and solutions. To make this happen, the committee acknowledged that both pre-service and in-service teachers will need on-going and deep professional development opportunities on climate change grounded on research-based approaches. Students will also need more opportunities to learn about the anthropogenic causes of current climate change to take bold action (Frame, 2020).

Although it feels as though this work should have begun long ago, ECCLPS leaders are acknowledging that climate change needs to be at the forefront of education now. Given the large scale of the problem, it is understandable that teachers might not know exactly what to teach. Where should you start integrating the science? How could you teach it in ways that empower young people, while also fulfilling the NGSS? What approaches are working for teachers that we can learn from? To help answer these questions and provide guidance to take on the challenge, I will draw upon the research of national and local educational agencies that focus on climate and environmental issues (including National Aeronautics and Space Administration [NASA], National Oceanic and Atmospheric Administration [NOAA], National Center for Science Education [NCSE], Yale Program on Climate Change Communication [YPCCC], and Aquarium of the Pacific to name a few). This chapter can seem technical and very detailed, but it provides teachers with the fundamentals of climate science to develop a shared language and launching point.

Public Attitudes Regarding Climate Change

According to on-going studies by Yale and George Mason University that began in 2008, there are six distinct groups that categorize Americans' beliefs, attitudes, policy preference, and behavior regarding climate change. Known as the Six Americas, these researchers discovered six audiences that respond to climate change information differently. The Six Americas are classified as Alarmed, Concerned, Cautious, Disengaged, Doubtful or Dismissive (Howe, Mildenberger, Marlon, & Leiserowitz, 2015). In 2020, the Alarmed and the Concerned constituted 54 percent

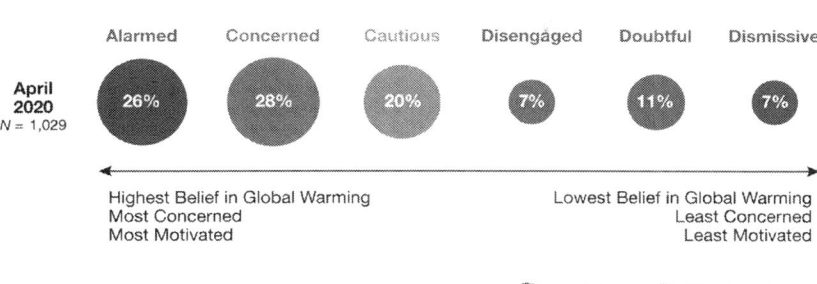

FIGURE 3.1 Yale's Report on the Six Americas
Image credit: Maibach, Leiserowitz, Roser-Renouf, & Mertz (2011). https://climate communication.yale.edu/about/projects/global-warmings-six-americas/

of the population while 25 percent of the population held low beliefs about the climate crisis (see Figure 3.1). The year 2020 marks the first time since Alarmists have now become the largest group in the Six Americas, and has tripled in size since 2014 (Goldberg et al., 2020).

In 2020, the YPCCC also found that 72 percent of Americans think that climate change *is* happening (see Figure 3.2), but only 57 percent understand that it's primarily caused by human activities. The public opinion mapping tool also revealed that 43 percent of the nation believes that climate change is harming them now (see Figure 3.3), while 71 percent believe it will harm future generations (see Figure 3.4). Lastly, 75 percent of Americans support regulating CO_2 as a pollutant, and 86 percent want to fund more research on renewable energy sources despite conflicting media messaging. What's clear is that the data reveals many differing views and opinions about climate change and what approaches should be taken. Understanding current climate is essential because among that population sample are science educators who are now responsible for teaching climate science under the new framework. It's important to know that the public at large is extremely supportive of students learning about climate science in schools.

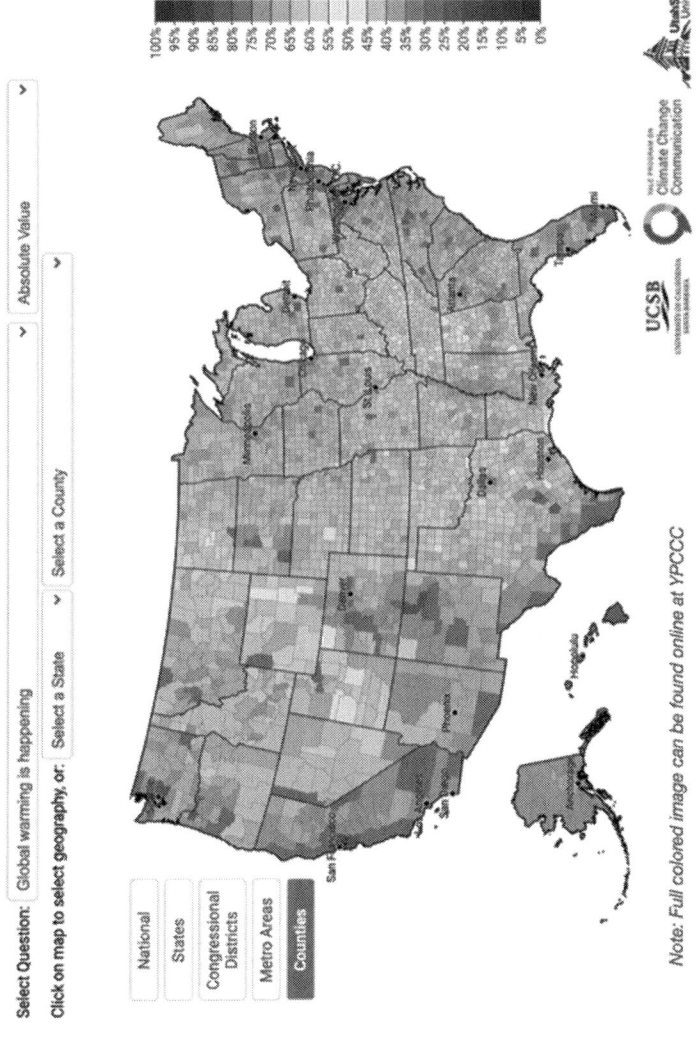

FIGURE 3.2 Yale Public Opinion Tool on Climate Change

Image credit: Howe, Mildenberg, Marlon, & Leiserowitz (2015) https://climatecommunication.yale.edu/about/
projects/global-warmings-six-americas/

Estimated % of adults who think global warming will harm them personally (43%), 2020

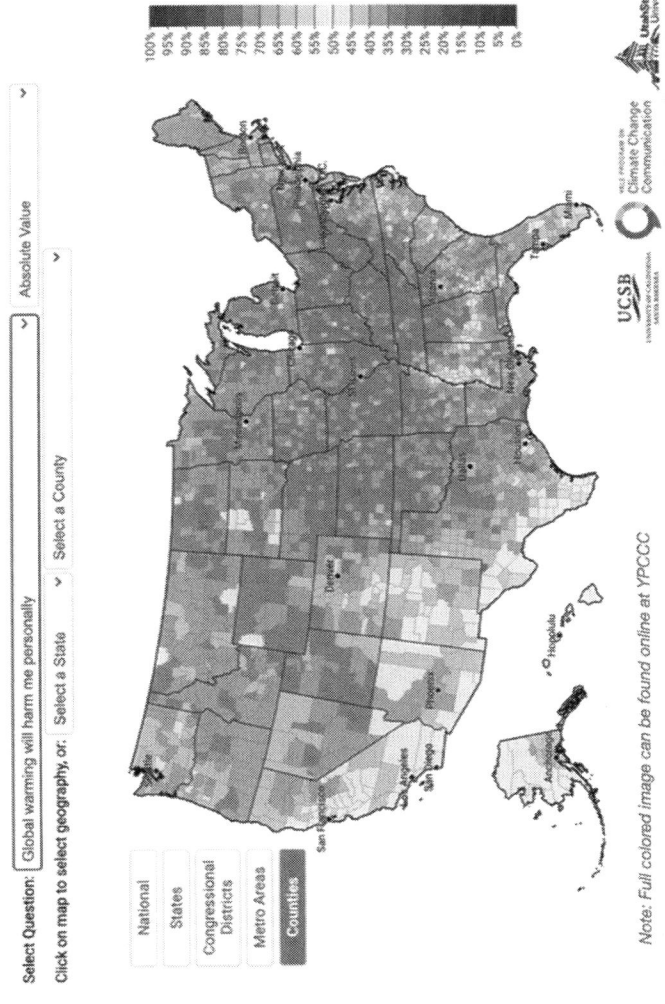

FIGURE 3.3 Yale Public Opinion Tool on Current Climate Impacts

Image credit: Howe, Mildenberger, Marlon, & Leiserowitz (2015) https://climatecommunication.yale.edu/about/projects/global-warmings-six-americas/

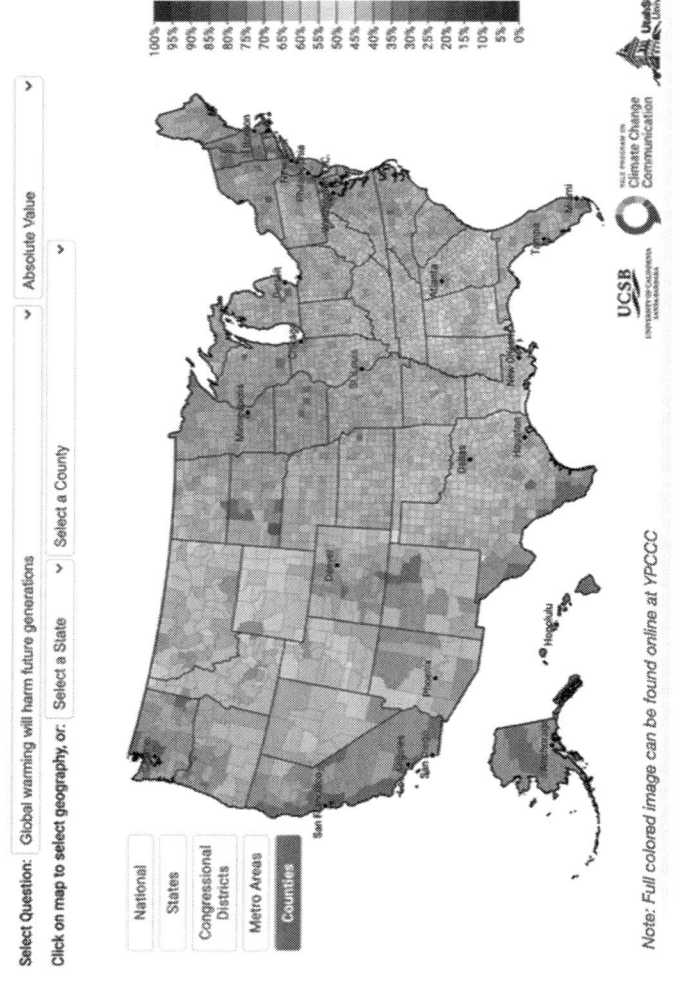

FIGURE 3.4 Yale Public Opinion Tool on Future Impacts of Climate Change

Image credit: Howe, Mildenberge, Marlon, & Leiserowitz (2015) https://climatecommunication.yale.edu/about/projects/global-warmings-six-americas/

 Consider taking and administering the Yale Six Americas Super Short Survey (SASSY) to see which audience you, your students, and their families identify with. Go to Bit.ly/YALEsassy and answer four short questions to get your results to compare their class data with national averages.

In looking at teacher attitudes regarding climate change, recall the 2016 study that revealed much confusion among science educators across the United States (Plutzer et al., 2016). The survey included 1,500 public secondary science educators and found that the median teacher allocates only 1–2 hours on climate change – covering nowhere near the depth of knowledge demanded by the NGSS. Roughly 52 percent of the teachers know the scientific consensus, and 68 percent of surveyed teachers understand that climate change is caused by human activities (Plutzer et al., 2016). Furthermore, the authors noted that nearly 50 percent of science educators did not learn about climate science during their undergraduate career. Therefore, more educational opportunities are needed to support teachers to effectively integrate climate science into their curriculum (Hestness et al., 2014; Lambert & Bleicher, 2013; Liu, Roehrig, Bhattacharya, & Varma, 2015; Plutzer et al., 2016; Sadler et al., 2004).

When looking at public opinions on whose responsibility it should be to educate students about climate change, YPCCC reveals that 78 percent of the public believe it needs to happen in schools (Howe et al., 2015). Figure 3.5 shows nationwide results of how people across all states agree that schools need to teach about climate change causes, consequences, and solutions. Teachers need educational support to teach the science of climate change, and it's helpful knowing that the general public supports (and expects) teachers to take on this role contrary to the conflicting messages broadcasted by the media.

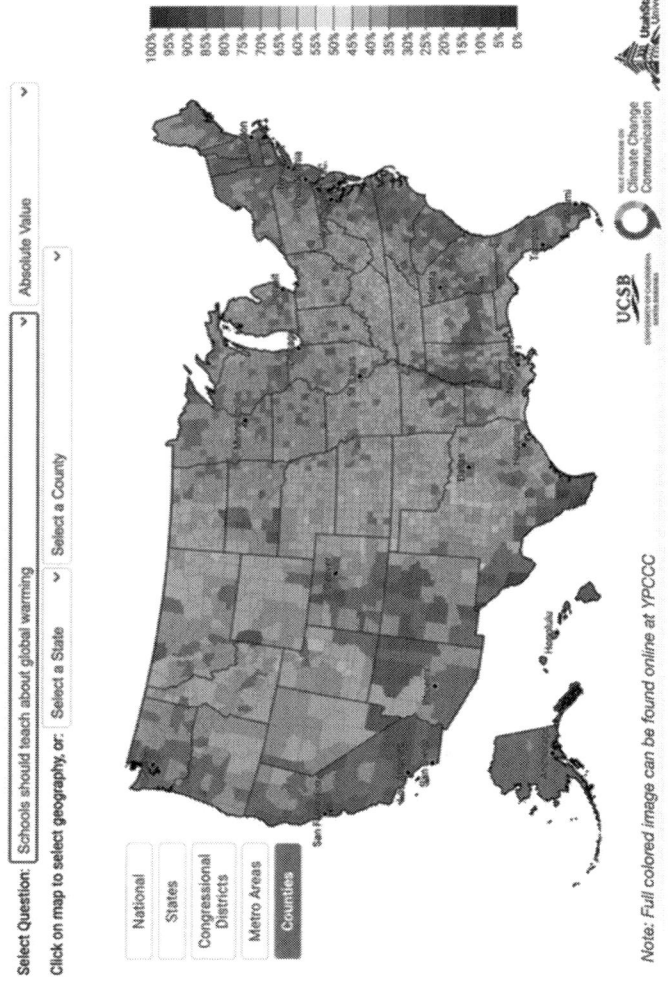

FIGURE 3.5 Yale Public Opinion Tool on Role of Teachers

Image credit: Howe, Mildenberger, Marlon, & Leiserowitz (2015) https://climatecommunication.yale.edu/about/projects/global-warmings-six-americas/

Identifying Your State of Mind

 As you reflect on this information, consider what your learning needs are regarding climate science and how you would potentially work with colleagues to build capacity at your school.

Your Current State

With a better understanding of public attitudes regarding the role of education and climate change, you can shift your focus to identifying learning goals you're ready to take on. We will refer to this state of mind as your current state. Acknowledging your current state is important because it means that you have reflected on your current theories of action, and recognize what is within your control as well as any potential challenges. The following activities in Exhibit 3.1 help to illustrate the possibility of change that can also be used with your students.

Exhibit 3.1 Activities to Illustrate Change

There are several activities that you can use to illustrate that change is possible. Consider taking some of these adapted activities back to the class when you integrate climate change content to position students as capable agents of change. *Note: There are many versions of these activities online to explore.*

Arms Crossed Activity

1. Ask students to stand up and cross their arms in front of them as though they are waiting for a friend to come out of class during a passing period.
2. Ask them to keep their arms there and take note of whether the left or right arm is on the top.
3. Ask them to release the hold, shake it off. Then cross again naturally and hold it there during discourse.
4. Ask students to share how that felt to just cross your arms (i.e. How did that feel? How easy was it to do that? Did

you have to think about which arm goes on top or on the bottom? How do you think you learned to do this?)

5. Now, ask them to release the hold, shake it off. Then cross again but this time, ask them to purposely put the opposite arm on the top.

6. Ask students to share how that felt (i.e. How did that feel? How easy was it to do that? Did you have to think about which arm goes on top or on the bottom? How do you think you learned to do this?)

7. As they stand there awkwardly holding this seemingly unnatural pose, facilitate a quick discussion about the possibility of change. Although it was weird to push students to control which arm goes on top, **they did it**. That is the point. Climate change is exacerbated because we take for granted that our everyday activities have a larger impact and although it would be "awkward," "weird," "uncomfortable," or even "difficult" to change our actions, it is important to note that it is **possible**. With enough practice over time, it would become second nature to do things differently.

Sustainability Activity

1. Hold up two pens for students to examine (make sure one pen is a disposable pen and the other is a fancier-looking pen).

2. Let students know that one pen costs about 10 cents and the other is a rare $100 collectable pen that you can buy ink cartridges for (it's okay to embellish because they will soon understand your point).

3. Ask them to explain to an elbow partner which pen would be considered more "sustainable" and "why they think that."

4. After facilitating this conversation and hearing from students or taking a class vote, have a discussion about sustainability and recognizing our collective relationship with material items.

5. The disposable pen may seem like the obvious choice to label as unsustainable because it's cheap and meant to be thrown away. It's important to be explicit about this and to also share that with these pens, you use them daily and rely on them for grading.

6. Most students would automatically say that the fancier pen is more sustainable because of its features. Explain that the fancier pen, however, was a gift given to you that is used once every few years when writing important letters. The pen is made from rare metals and the ink cartridge is expensive to replace. Ask students to rethink "sustainability" when relationships are considered.

7. The problem most students might not realize is that even if the fancier, more expensive pen is more sustainable, we choose not to use it very often and therefore our relationship with that item makes it unsustainable. If we're constantly opting for the cheaper disposable option because we have a comfortable and reliant relationship with those items, that is a problem. We need to change our relationship with the items we rely on in order to push ourselves to change. This activity also works well with water bottles and canteens as well (people opt for disposable water bottles every day because they think the canteen is too nice to use, keep forgetting to fill it up with water, etc.).

 Consider bringing these activities back to the class and share your experiences with the network at www. empoweredscience teachers.com (Book Resources → Chapter 3 → Discussion Board)

Determining Your Desired State

Reflect on the current state of climate education and public attitudes regarding climate science. To what degree is your school, department, or community addressing climate change? Do your answers align with the YPCCC public opinion maps presented earlier? What vision do you have for the future of science education? The next chapter will support your continued growth in launching storylines but remember to plan with the end in mind. Think about your desired state, and what steps will be needed to get there. What will you need to accomplish with your students and colleagues as you take on climate change? To help organize the content in this chapter, choose a curriculum map template to help create storylines that follow your school and district parameters (see Appendix B). Please note that the term "units" will be used interchangeably with "instructional segment(s)" in this chapter to begin integrating vocabulary used in the NGSS. First, access your school's science curriculum map (if any) for major units or topics that you are required to cover for NGSS. This might be something you have little control over, but important to seek out. You might not be able to control the flow of the units, but you can control how the segments are taught in your class. You can also center your curriculum around climate change, environmental literacy, human impact, or impact on humans to fulfill the NGSS (and Environmental Principles and Concepts for Californians).

Continue reading this after you have written out the skeletal structure of your curriculum map (identifying all the major topics for your subject area/grade level), it's time to add new ideas and concepts to your map. One way to do so is to add post-it notes of essential climate change content, any supporting resources, and phenomena that might work well in that instructional segment. The next segment will provide key ideas of climate science, and the additional teacher resources at the end of the chapter will provide rich phenomena to launch your units. Once you feel that you have included enough, review all the post-it notes in that category to see how they can flow

 Before you continue to the next segment, go online to www.empoweredscienceteachers.com to select a blank curriculum map to use to organize the following content, ideas, or resources.

Climate Change Is Complex ◆ 85

from one idea to another and from concept to concept. Do you notice immediately which concepts are revisited across topics? Can you identify how ideas expand each time you teach about them across your instructional segments? Lastly, take a look at all the phenomena you organized for one instructional segment and identify *one* anchoring phenomenon big enough to connect all the ideas together (directly or indirectly). See Appendix C for iterative examples provided by other teachers from the climate change educational program.

Climate Science Fundamentals

To understand the basics of climate science, it is essential to understand the difference between climate and weather, climate in relation to Earth's systems, and how climate has changed due to both natural forces and human activities. Climate refers to average weather conditions (precipitation, temperature, wind, etc.) with consideration for extremes that regions may experience throughout the year. Any fluctuation in temperature, rainfall, snowfall, or wind that only lasts for hours, days, or weeks is considered weather. Earth's average surface temperatures allow for an abundance of water and the thin layer of atmospheric gases keeps our planet warm enough to sustain life. Refer to Table 3.1 to get clarification on common terms used in climate science. A common analogy is thinking about your entire wardrobe as climate but your daily clothing choices as the weather. You might have outfits for when weather deviates, but your overall wardrobe caters to the climate of where you live.

Earth's Energy Budget

In my interview with a senior specialist at NASA, I wanted to know the most important concepts to teach when addressing climate change in science classrooms. The first thing that came up was teaching about Earth's Energy Budget. Understanding how energy from the sun flows in and out of Earth's atmosphere,

TABLE 3.1 Clarification on Terms

Vocabulary	Description
Anthropocene	The geological age where human activities are the dominant influence on the environment and Earth's systems.
Anthropogenic	Resulting from human actions or activities.
Atmosphere	The layer of gases surrounding a planet.
Biosphere	All living and nonliving organisms on Earth.
Carbon Cycle	The cycling of carbon between Earth's systems.
Climate	The average weather conditions (precipitation, temperature, wind, etc.) with consideration for extremes that regions may experience throughout the year.
Climate Change	The extra heat energy from the increased greenhouse gases causes other changes too (such as sea level rise, changing rain patterns, collapse of ecosystems, ocean acidification, larger than ever storms, etc.), that can have large impacts on all of Earth's systems. This term captures all those changes as a result of global warming.
Geosphere	Referring to the solid Earth.
Global Warming	An increase in global temperatures, which is caused by an increased greenhouse effect that traps more heat energy in the Earth's systems (Compare with Climate Change).
Hydrosphere	Referring to all water on Earth (oceans, lakes, rivers, streams, water vapor).
Thermohaline Circulation (Earth's Great Conveyor Belt)	Cold dense North Atlantic sea water circulates between warmer continents near the equator causing water to rise to the surface (upwelling). The water continues to travel to Antarctica where it cools, becomes more dense, and sinks again as it circulates back to the North Atlantic sea. Note that the warmer waters lose their nutrients, so this cycling supports marine life by bringing nutrients to the bottom of the ocean as colder – more dense waters.
Weather	Any fluctuation in temperature, rainfall, snowfall, or wind that only lasts for hours, days, or weeks.

where the energy travels, what happens to it, and seeing how Earth balances the radiant energy are all essential to understanding climate science. Earth's systems are constantly trying to balance the energy received from the sun to reach equilibrium. This occurs when natural (such as volcanic eruptions) and anthropogenic phenomena (such as burning fossil fuels) release energy and super greenhouse gases (GHG) into the atmosphere. As a result, temperatures increase or decrease to reach a state of equilibrium.

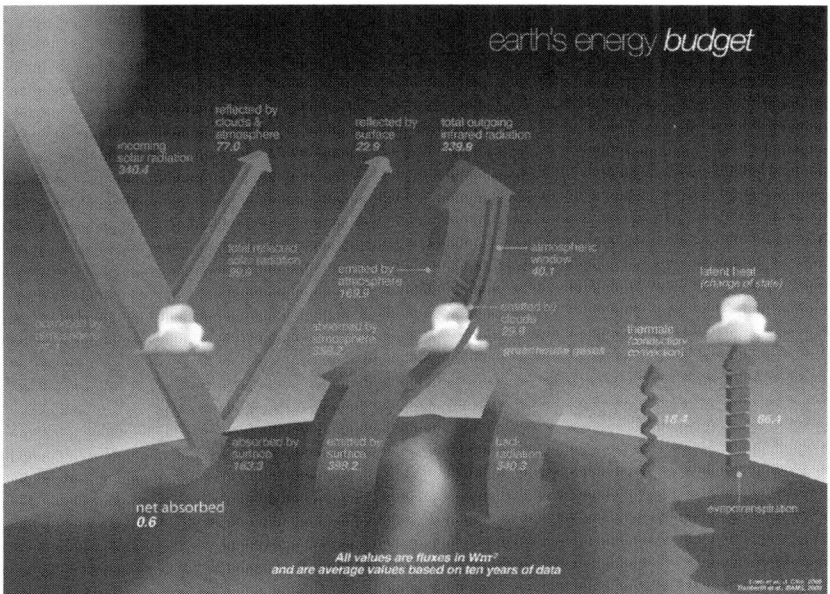

FIGURE 3.6 NASA's Visual on Earth's Energy Budget
Image credit: NASA

There are four major paths that the sun's energy can take, which may or may not result in heat radiating back to space (see Figure 3.6 provided by NASA to see the Earth's energy budget).

- ◆ Path 1: Rays enter the atmosphere and are absorbed in the atmosphere or clouds. Some of the heat radiates back to space.
- ◆ Path 2: Sunlight directly hits the ground and heats the ground surface. Some of the heat radiates back to space.
- ◆ Path 3: Rays directly reflect back to space by reflecting off the atmosphere, clouds, or the ground.
- ◆ Path 4: It's important to note that even when the sun's rays cannot be seen during nighttime, heat continuously radiates out to space causing the environment to become colder.

Recall that learning about the Sun's pathways will allow students to understand how some energy does not radiate back to space causing the planet to trap excess energy resulting from excess GHG emissions.

The gases in the atmosphere act like a heat-trapping blanket wrapped around the Earth. When we produce excess green-house gases (such as carbon dioxide, nitrous oxide, methane, etc.), human activities thicken the blanket and as a result trap more heat. Imagine lying in bed under the covers and breathing under the blanket. Eventually you will increase the tempera-ture under the blanket, and it will become hot causing you to want to cool down by taking the blanket off. As we increase the heat in the atmosphere due to anthropogenic factors, Earth is trying to balance the excess hot temperatures causing a variety of changes throughout Earth's systems (such as melting ice caps, sea-level rise, ocean acid-ification, higher heat indexes and humidity levels, and much more). The difference is that Earth is overheating without a clear way to "remove the blanket" for all the heat trapped underneath.

 Take it back to the class by utilizing other evidence-based communication strategies from the National Network for Ocean and Climate Change Interpretation (NNOCCI). Go to Bit. ly/NNOCCIcards to use their reframe cards.

Global Climate Change

Although skeptics argue that climate has always changed and that the planet is going through a period of warming due to natural phenomena such as volcanic activity, scientific research reveals, however, that the current rate of change is ten times faster due to human activities. If we measure Earth's average temperature beginning roughly 9000 Before the Common Era (BCE), tempera-ture reached modern levels and remained at roughly between 0°C and 1°C. It stays in this range and dips down to roughly -1°C and 0°C between 1CE and the 1850s at the start of the Industrial Revolution. With the introduction of fossil fuels, carbon dioxide

(CO_2) emissions begin to rapidly increase in the atmosphere along with other super GHG. The start of The Great Acceleration of the 1950s began to shift Earth's average temperatures. It is now trending towards 1°C, and our current predicted path is an increased 4°C by the year 2100 if we don't intervene. With that climate model, we are looking at the collapse of ecosystems, endangerment of marine species, sea level rise threatening coastlines, more extreme weather events, impacts on agriculture, and much more. Figure 3.7 is a visual showing the timeline of Earth's average temperatures to help people understand the significant increased rate of change over the centuries. As you explore the visual showing the drastic shift in temperatures since the Industrial Revolution, this is a great way to debunk the misconception that climate has always changed and that humans have not contributed to the current increase in greenhouse gases.

Earth's Systems

The Earth's climate should also be understood as a complex system where many parts interact with one another to create climate conditions. We are focusing on four primary systems that make up the Earth's systems often referred to as the atmosphere, hydrosphere, geosphere, and the biosphere. The atmosphere refers to an invisible blanket of gases surrounding the Earth that contains gases such as nitrogen and oxygen, and smaller amounts of trace gases such as water, carbon dioxide, methane, and ozone. All of these gases influence Earth's climate system regardless of their small quantities. The hydrosphere refers to the water on Earth's surface and atmosphere (in liquid, solid, or gas form). Water stores a great deal of heat and absorbs large amounts of CO_2. The oceans circulate the heat around the globe through ocean currents, which transfers energy to the atmosphere playing a large role in influencing Earth's climate. The geosphere refers to Earth's land from the surface down to the core. Rocks play a large role in shaping Earth's climate because of how much carbon they can store, and volcanic eruptions release a great deal of particles and gas into the air impacting Earth's climate. Lastly,

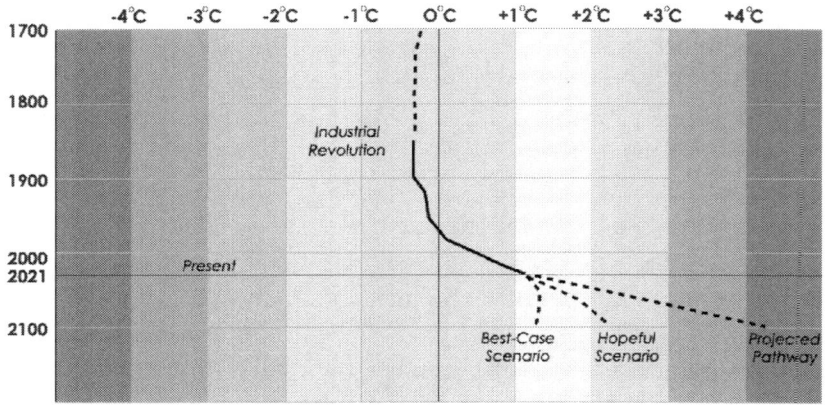

FIGURE 3.7 Earth's Average Temperatures

the biosphere refers to all living things on Earth. Living things impact each system because of the amount of carbon they can emit or absorb. Living things are also impacted by the changing climate as we learn more about the sixth-mass extinction and the resulting destruction or collapse of ecosystems. Ultimately, Earth's systems are complex in that many interactions occur between systems as they influence each other and the climate.

Multiple Carbon Cycles

Teachers who teach about carbon cycles generally do so from specific perspectives (life or physical). There are multiple perspectives of the carbon cycle that occur because carbon moves through each of the Earth's systems in different ways. When individuals fully understand the impacts of their daily decisions, they begin to care about and take action on the problem. The carbon cycles can help students understand the source and impacts of excess anthropogenic emissions that are causing ocean acidification, dissolving of the sea floor, damaging key marine species like phytoplankton, melting of the permafrost, among many other harmful impacts. This segment will provide an overview of the various carbon cycles to help you see the larger picture.

The Carbon-14 Dating Cycle (a Physical Perspective)

The Carbon-14 (C-14) dating cycle allows for students to understand how carbon enters living things and how it could be released due to anthropogenic activities (such as burning fossil fuels or drilling for natural gas). Scientists can use C-14 radioactive dating to figure out how old something is based on when it died. They do so by typically analyzing elemental half-lives in remaining bone, wood, or cloth samples. This process starts with the sun's rays colliding with atoms in the atmosphere to create high energy neutrons and eventually moving to living species consuming the carbon-rich nutrients. Towards the end, the cycle is complete when decomposers work to put nitrogen back into the atmosphere from decayed matter (see Figure 3.8 for more details). As living things decay and are buried over time, they store carbon that remains virtually undisturbed (with exceptions of both natural phenomena and human activities).

The Marine Carbon Cycle

The marine carbon cycle has three paths that interact with each other allowing some carbon to return to the atmosphere. This cycle helps students to understand how the ocean absorbs at least a quarter of the carbon released (in the form of CO_2), and acts as a circulatory system for Earth. The ocean is also known as Earth's "heart" because it helps the planet to reach equilibrium by absorbing excess heat in the atmosphere. As the ocean cycles heat and nutrients across the planet through thermohaline circulation (the great ocean conveyor belt), it works as a regulatory system for Earth.

Path 1: Carbon dioxide from the atmosphere dissolves into the ocean and sinks to lower depths with colder waters. Through thermohaline circulation, the CO_2 eventually rises up as bubbles with warmer waters due to crashing ocean waves and returns to the atmosphere. This process is slow and relies on the turning of the ocean waters as it travels from the equator towards the poles (view NASA's animation at Bit.ly/NASATC). Scientists have been warning us that the conveyor belt is beginning to slow and weaken due to climate change. These are impacts we

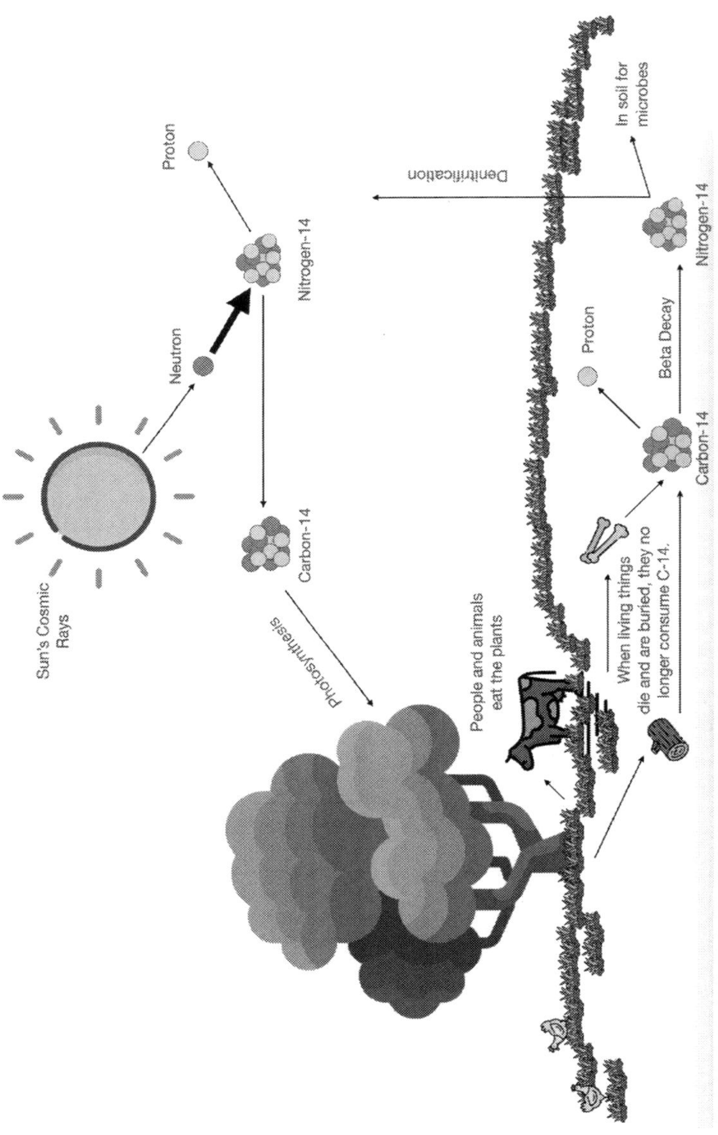

FIGURE 3.8 Carbon-14 Radioactive Dating Cycle

are currently seeing and feeling through colder winters, hotter summers, direct impact on water and food supply, and more.

Path 2: Carbon dioxide from the atmosphere dissolves into the ocean and is absorbed into phytoplankton and algae as they engage in the process of photosynthesis (similar to how plants do this on land). Through the food chain, some of the carbon returns to the atmosphere when sea creatures participate in respiration. When sea creatures die in the ocean and sink to the bottom of the ocean they decay. The bacteria begin to decompose the dead sea creatures and breathe out CO_2 that returns carbon to the atmosphere in the process.

Fossil fuels (like oil and natural gas) formed in the ocean are the result of dead marine species (such as plankton) that have been buried over time by mud and sand. The remains change into oil due to high pressure, high temperatures, and the help of bacteria. The remains can also change into limestone, which is commonly extracted to make concrete. These non-renewable sources of energy take millions of years to develop under very specific conditions. This carbon would normally be stored underground, but humans have been extracting fossil fuels for energy. That means that when we burn fossil fuels (which is anthropogenic), we are releasing excess amounts of carbon into the air that otherwise would have been stored underground.

Path 3: This path explains ocean acidification and how the ocean absorbs excess CO_2. The CO_2 in the atmosphere dissolves in the ocean producing carbonic acid (H_2CO_3). Due to the nature of acids, carbonic acids can lose a hydrogen-ion when reacting with water, which converts it into bicarbonate-ion (HCO_3^-). When bicarbonate-ions (which are acidic) lose a hydrogen-ion it turns into carbonate-ions (CO_3^{2-}). Marine species (such as lobsters, crabs, phytoplankton, etc.) rely on carbonate-ions to create their shells.

There are equilibrium reactions that occur between these three chemical reactions as well (see Figure 3.9). If there are lower levels of CO_2 in the atmosphere compared to the level in the ocean, carbonic acids will release carbon back into the atmosphere and the chain reaction described previously occurs. On the other hand, if there are increased levels of carbon dioxide in the atmosphere

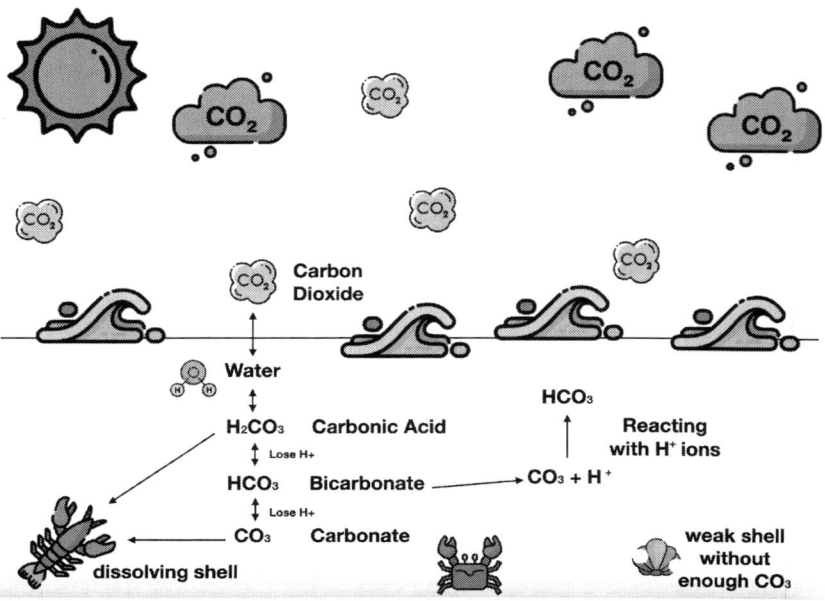

FIGURE 3.9 Carbon Cycle and Ocean Acidification

then more carbon is dissolved into the ocean forming more carbonic acid, bicarbonate-ions, and carbonate-ions. The carbonate-ions in the water continue to react with dissolved CO_2 to form more bicarbonate-ions, which allows the ocean to hold ten times more carbon dioxide (known as the ocean buffer). Remember that carbonate-ions are needed for marine species, but when carbonate reacts with carbon dioxide it produces bicarbonate-ions (which cannot be used for shell production). As a result, shells formed are weaker without enough carbonate-ions, and the current shells begin to dissolve because of the carbonic acid in the ocean. This is known as ocean acidification (see NOAA for more Bit.ly/NOAAoceanacid).

The Geologic Carbon Cycle
Another carbon cycle to consider occurs in the geosphere, where volcanic eruptions or metamorphic events release CO_2 into the atmosphere naturally. Rocks in the ground play a big role in storing carbon and other gases through the rock cycle (access

NASA's resource at Bit.ly/NASAROCK). Similar to the marine carbon cycle where CO_2 reacts with water to form carbonic acid, this happens when rain water comes into contact with rock sediment, which starts the chemical weathering process. As rivers carry the various ions into different bodies of water (lakes, streams, oceans, etc.) a chemical reaction occurs. The sediment builds up over time and buries any dead marine specimen over time. Layers continue to build through this process, and the increased pressure turns shells and sediment into limestone allowing for carbon capture.

The Anthropocene

Now that you understand the natural and anthropogenic phenomena taking place through the energy budget and various carbon cycles, how do we know what the cause of current climate change is? Evidence shows that current climate change is due to humans. We are such a force of nature as a species that we are able to collectively alter every Earth system with our actions (or inaction). Recall that this geological time period marked by human actions is known as the Anthropocene. "The Age of Humans" is a result of the Industrial Revolution, Great Acceleration, and dramatic population growth (Ramanathan et al., 2019). These advancements have allowed humans to thrive and make technological advances for a better life on Earth, but we have done so at the expense of everything else (Table 3.2).

Current climate change is anthropogenic because the excess GHG emissions are a direct result of human activities and actions. One way to help students understand anthropogenic climate change is to have them compare regular and rampant CO_2 emissions. The NGSS calls for teachers across science subjects and grade levels to teach the carbon cycle pathways. This is one way to showcase natural events that release carbon dioxide into the atmosphere compared to the excess levels of GHG that humans emit due to daily activities. It may be shocking for students to learn that humans emit almost 100 times more carbon dioxide than all volcano eruptions combined, so the recent increase in

TABLE 3.2 Teaching about the Anthropocene

Topic	Description
Climate Reality Project	Check out this resource for ways to teach students about anthropogenic climate change. Link: Bit.ly/CRPHUMANS
UC-CSU NXTerra – Transformative Education for Climate Action	Review a variety of resources that provide education, tools, and teaching tips on approaching the climate crisis. Link: Bit.ly/TEACHINGGCC
NASA – Images of Change	Have students analyze images from across the globe on the impacts of climate change. Link: Bit.ly/PicChange
Our Climate Our Future	Explore teaching resources and student videos to introduce the causes of climate change and ways to get involved through the Alliance for Climate Education. Link: https://ourclimateourfuture.org/
Specific Topics to Explore for Concrete Examples of Anthropogenic Factors	Climate change linked to consumerism. Production and the role of plastics (such as plastic bottles, bags, take out containers, etc.). Deforestation, reliance on palm oil, and rise of diseases. Burning Fossil Fuels. Extraction and uses of limestone. Travel (commuting for work or for leisure) Food production and transportation. The raising and consumption of cows, pigs, chickens, etc. Deep sea ocean or permafrost drilling. The role and effects of fracking.

temperatures as a result of excess CO_2 cannot be attributed to natural phenomena (Ramanathan et al., 2019). If people do not learn climate science and fail to see how scientific data supports the claim that humans are the cause, they will not feel the urgency of this crisis or take the necessary actions needed to bend the curve. *Project Drawdown* has an amazing video series to go through the climate science as well (Drawdown.org/climate-solutions-101). See Table 3.2 for additional resources on how to teach anthropogenic climate change.

Scientific Consensus on Climate Change

Although the NGSS requires secondary science educators to teach climate change content, many teachers report avoiding the topic because they considered it highly controversial and heavily debated among the scientific community (Hestness et al., 2011; Liftig, 2012; Maibach et al., 2014). When science educators receive climate change information from media sources, the information may not reflect scientific research or may portray climate science as controversial. Maibach et al. (2014) found that 42 percent of Americans believe that most scientists think that climate change is happening, and 33 percent believe that there is no scientific consensus. This is problematic when 97 percent of the scientific community agrees that climate change is occurring due to human-related activities (Cook et al., 2013, Cook et al., 2016). There's actually *no* debate about whether climate change is happening among scientists. Instead, a small number of scientists (less than 3 percent and many in fields outside of climate science) are questioning whether it is due to anthropogenic causes because the data collection is on-going. Teachers should not have students debate about climate change because it sends the message that personal opinions outweigh credible data gathered by climate experts. Debates should instead be about policy or ethical dimensions of climate change (such as the growing number of climate refugees, the growing food crisis, what to do about the collapse of ecosystems, etc.), but not about the science of climate change.

 Take it back to the class with the following resources to introduce scientific consensus to students to debunk misconceptions circulating through the media.

1. Go to Bit.ly/CookConsensus for teacher resources provided by Skeptical Science.
2. Show students John Oliver's Mathematically Representative Climate Change Debate – Bit.ly/CCDEBATE
3. Have students analyze Cook's article for data on consensus – Bit.ly/COOKconsensus
4. Use NCSE's teaching resource to counter skepticism – Bit.ly/NCSEscepticism

Major Teaching Takeaways

Although the following five points are layered with a great deal of content, these are the major teaching takeaways for any science teacher looking to teach climate change.

1. **Teach the scientific consensus on climate change** – This is the perfect opportunity to teach the Nature of Science, argumentation in science using credible evidence to support a claim, and provide clarification for students on how this topic is politically controversial but not scientifically.

2. **Do not debate about climate change** – Allowing for students to debate about climate change sends the message that their opinion trumps credible scientific evidence and data in the end. Climate change debates should focus on the ethical or political dimensions of the climate crisis after learning about scientific consensus. (For example, should we continue to rely on palm oil for our low-priced processed foods or try to find another solution that protects the world's rainforests?)

3. **Teach climate change through the lens of systems thinking** – Consistently come back to the crisis throughout different segments and lessons to show how interconnected science and the impacts of climate change are (e.g. Direct impacts on the hydrosphere also impact the other spheres, which have rippling effects, everyday plastics end up in the ocean, etc.). Remember that concepts or ideas not revisited often will not be retained or transferred by students. The NGSS also stresses teaching science through systems-thinking to show the cyclical and iterative nature of science.

4. **Teach climate change as a socioscientific issue** – Research reveals that this approach to climate science is

effective in helping students to develop scientific literacy skills by learning more about the Nature of Science. To do so, (1) Consider how you design your lessons around climate change phenomena, (2) be open and willing to learn and teach about a complex subject, and (3) integrate opportunities to address the ethical and social dimensions related to the topic because it directly impacts students and has the ability to empower students as agents of change (For example, students could analyze data and pose higher-level questions using vetted online tools when seeing direct community impacts).

5. **Teach the cause of current climate change along with the need to seek diverse and intersectional solutions.** – The NGSS emphasizes the need to integrate "human impact" through all sciences in iterative ways. As teachers take on education for climate action, they first need to teach about the anthropogenic causes. Put simply, if humans are the cause then we're also the answer. Students will want to know what they can do to help adapt or mitigate climate change and understanding that their direct actions are tied to the climate crisis might influence their daily decisions.

As you consider what science content to include or exclude in your existing or newly developing curricula, think about how you can anchor storylines around climate change. This will allow you to build on students' interests, their lived experiences, funds of knowledge, critical thinking skills, and so much more. Recall from previous chapters that incorporating the content is one way to address the NGSS framework, but the complementing piece is thinking through the delivery of that lesson to build students' capacity as future leaders and problem solvers that can apply what they have learned. The next chapter will go through more pedagogical practices and resources to successfully anchor your curriculum on climate change.

Collective Voices for Climate Change Education
Katie Kozma (Instructor at Reef Check Foundation)

I think the most important thing to teach students about when it comes to climate change, is that there ARE things that they can do in their everyday lives to try to reduce their carbon footprint and try to reverse the effects that past generations have put on our home planet. Being educated about what's happening and what's predicted to happen if we keep going down the same path is the first step. Knowledge is power. Students need to be very aware of the actual reality of the effects that we pose in our planet's future. Teachers need to try to **empower** *students to get out and spread the word about solutions to the problems we're currently facing and ways that they can help change the course of what's to come. I think that a big issue we come across right now, is the lack of hope that a change can be made at this point. We need to put out into the world the idea that one person* **can** *change the course of the world. If that one person makes a change and spreads the word about that change they have made, this can ultimately encourage others to join the fight to defeat the climate crisis.*

Carol Ann Hagele (Education Specialist at SGV Mosquito & Vector Control District)

I teach mosquito-borne disease awareness in schools because humans are not even close to winning the battle against mosquitoes. **Temperature** *plays a big role in mosquito activity. Humans are facilitators of the spread of dangerous invasive mosquitoes and provide habitat. But it is temperature that determines the speed of mosquito growth and replication of the viral pathogens they transmit, and how long each year transmission occurs.*

Climate change drives many of the challenges to global public health. Mosquitoes are just one aspect of those changes. Students need to understand that **humans** *are a powerful ecological force and that there are overarching ecological rules that must be followed to live on this planet. Several hundred years ago, humans found workarounds or cheats to ecosystem rules. We are now realizing that planetary system-limits will eclipse any ecological cheats humans can come up with. It is important to facilitate a deeper understanding in students of how this planet functions and how to preserve and maintain the life-sustaining ecosystems already in place if they are to create a path to a healthy future.*

Additional Teacher Resources

Access "Bending the Curve" to learn about climate solutions –
   Bit.ly/BTCBOOK
Access TedEd United Nations online environmental lessons –
   ed.ted.com/earth-school
Curious if your state is "Making the Grade" on
   climate education? –
   climategrades.org/
Check out HHMI's Geologic Carbon Cycle resources –
   Bit.ly/HHMICarbon
Check out NASA Teacher Resources –
   bit.ly/TeacherWL
Explore the NxTerra website for climate change resources –
   Bit.ly/NXTerra
Get project-based curricula on environmental topics from SEI –
   Bit.ly/TEACHSEI
Help students comb through fake-news as fact-finders –
   Bit.ly/FactDetectives
Learn about the Anthropocene here –
   www.anthropocene.info/
Look over the NCSE website for teacher resources –
   ncse.ngo/teaching-climate-change

Research showing the flaws of climate skeptics' research –
   Bit.ly/StudyFlaws
Take a look at the UCLA ECCLPS program –
   Bit.ly/ECCLPS
Teach using the full visual of Earth's Average Temperatures by
   XKCD –
   Bit.ly/XKCD1732
Use the CalAdapt tool to see community impacts –
   cal-adapt.org/
Use this tool to see the real-world cost of climate change –
   www.impactlab.org/
View Various Yale Climate Opinion Maps 2019 –
   Bit.ly/YaleMaps
Watch "Before the Flood" documentary –
   www.beforetheflood.com/
Watch "Chasing Coral" documentary –
   www.chasingcoral.com/
Watch John Oliver's debating climate change clip –
   Bit.ly/JODEBATE
Watch "Racing Extinction" documentary –
   racingextinction.com/

References

Cook, J., Nuccitelli, D., Green, S. A., Richardson, M., Winkler, B., Painting, R., ... & Skuce, A. (2013). Quantifying the consensus on anthropogenic global warming in the scientific literature. *Environmental Research Letters*, 8(2), 024024.

Cook, J., Oreskes, N., Doran, P. T., Anderegg, W. R., Verheggen, B., Maibach, E. W., ... & Nuccitelli, D. (2016). Consensus on consensus: a synthesis of consensus estimates on human-caused global warming. *Environmental Research Letters*, 11(4), 048002.

Frame, A. (2020, April 21). *Climate Change: It's Not One More Thing – It's the Thing*. Retrieved October 15, 2020, from https://tenstrands.org/ci/climate-change-its-not-one-more-thing-its-the-thing/

Goldberg, M., Gustafson, A., Rosenthal, S., Kotcher, J., Maibach, E., and Leiserowitz, A. (2020). *For the first time, the Alarmed are now the largest of Global Warming's Six Americas*. Yale University and George Mason University. New Haven, CT: Yale Program on Climate Change Communication.

Hestness, E., McDonald, R. C., Breslyn, W., McGinnis, J. R., & Mouza, C. (2014). Science teacher professional development in climate change education informed by the next generation science standards. *Journal of Geoscience Education*, 62(3), 319–329.

Hestness, E., Randy McGinnis, J., Riedinger, K., & Marbach-Ad, G. (2011). A study of teacher candidates' experiences investigating global climate change within an elementary science methods course. *Journal of Science Teacher Education*, 22(4), 351–369.

Howe, P., Mildenberger, M., Marlon, J., & Leiserowitz, A. (2015). Geographic variation in opinions on climate change at state and local scales in the USA. *Nature Climate Change*. DOI: 10.1038/nclimate2583.

Lambert, J. L., & Bleicher, R. E. (2013). Climate change in the pre-service teacher's mind. *Journal of Science Teacher Education*, 24(6), 999–1022.

Liftig, I. (2012). A tough climate for teachers. *Science Scope*, 35(7), 1–1.

Liu, S., Roehrig, G., Bhattacharya, D., & Varma, K. (2015). In-service teachers' attitudes, knowledge and classroom teaching of global climate change. *Science Educator*, 24(1), 12.

Maibach, E. W., Leiserowitz, A., Roser-Renouf, C., & Mertz, C. K. (2011). Identifying like-minded audiences for global warming public engagement campaigns: An audience segmentation analysis and tool development. *PloS One*, 6(3), e17571.

Maibach, E., Myers, T., & Leiserowitz, A. (2014). Climate scientists need to set the record straight: There is a scientific consensus that human-caused climate change is happening. *Earth's Future*, 2(5), 295–298.

Plutzer, E., McCaffrey, M., Hannah, A. L., Rosenau, J., Berbeco, M., & Reid, A. H. (2016). Climate confusion among US teachers. *Science*, 351(6274), 664–665.

Ramanathan, V., Aines, R., Auffhammer, M., Barth, M., Cole, J., Forman, F., et al. (2019). *Bending the curve: Climate change solutions*. California: Regents of the University of California. Retrieved from https://escholarship.org/uc/item/6kr8p5rq

Sadler, T. D., Chambers, F. W., & Zeidler, D. L. (2004). Student conceptualizations of the nature of science in response to a socioscientific issue. *International Journal of Science Education*, 26(4), 387–409.

4

Climate Change as the Anchor

Read this when:

- ◆ *You understand the fundamentals of climate science and are ready to integrate the content into your curriculum.*
- ◆ *You want to learn more about how to develop students' scientific literacy skills.*
- ◆ *You want to learn about phenomena-based instruction to begin designing storylines.*

Lessons from Bonsai Koi Fish

Did you know that bonsai koi fish that live in healthy natural ponds can grow up to two feet in length, but those that grow in man-made tanks will not outgrow the container they live in? The most notable effect is that confined koi fish experience stunted growth due to environmental factors that would otherwise allow for them to grow much larger. As a result, they also have shorter life spans than those that live in healthy natural ponds with thriving ecosystems. Pause here to think about our educational system. Who gets to determine what and how much students learn or don't learn? Are we willing to leverage our decision-making power in the classroom to disrupt traditional science education so that students learn 21st Century skills for the problems they face in their communities now? I ask once again, if education was created and written to benefit one group of people up until the Civil Rights Movements, how can we revamp it to support every learner as NGSS and our society demands to combat urgent issues such as the climate crisis?

It's no secret that students rely heavily on their teachers as trusted sources for data and information. Don't forget that teachers hold a great deal of power in the classroom as individuals who determine what will be taught, how it will be taught, how students will be positioned, what is deemed worthy of time, and so much more. Given this important role, it is essential to support teachers with the paradigm and pedagogical shifts needed to teach socioscientific issues such as climate change.

There is a vast amount of research that reveals how highly dependent people are on the media and Internet to learn about global climate change (Caranto & Pitpitunge, 2015; Carter & Wiles, 2014; Hansen, 2010; Hestness et al., 2014; Hodson, 2003; Matkins & Bell, 2007; Somerville & Hassol, 2011). Among these individuals are teachers and students who also rely heavily on social media to get scientific information. Remember that climate change is an SSI because students are bombarded with messages and claims about it, whether teachers actively address it or not. What's clear is that when teachers choose to avoid or omit climate change altogether, they are indirectly sending messages to students about their underlying core values and beliefs (Kolstø, 2001; Sadler et al., 2004). Just as antiracist science teaching requires an active and intentional effort rather than choosing the safety of silence (Kendi, 2019), teaching about climate science requires a deliberate effort. This is an opportunity to engage students in learning about the science of climate change to provide them with a safe space to ask questions, debunk misinformation, express their emotions, model life-long learning, use data and credible evidence to support their claims, engage in science and engineering practices, access their agency to take action on direct impacts to their communities, and so much more. Teachers are powerful agents of change, and these are great opportunities for students to experience uninhibited growth.

100Kin10 Predictions

100Kin10 is a national organization that seeks to provide students with high-quality STEM education through recruiting and retaining 100,000 excellent STEM teachers by 2021. Through their work and data collection, they have identified seven grand challenges underlying the STEM teacher shortage that require solutions. These include professional growth, teacher preparation, ensuring schools value the S, T, and E in STEM, and several others. In their 2019 Predictions Report, 100kin10 predicts that environmental advocacy will engage more students in STEM. Worldwide, students are taking part in movements and opportunities for activism that address environmental challenges such as climate change. Students are taking the lead and the data reveals that schools need to start responding to these growing interests to support these efforts.

The NGSS has the ability to catalyze climate change education, while enhancing scientific and environmental literacy efforts as an intersectional issue. This chapter will provide you with the tools, knowledge, and resources needed to tap into your students' agency. As the desire to mobilize on climate change increases among our nation's youth, teachers can support that growth and desire with purposeful curriculum design that centers on the climate crisis.

Asking More Questions

In Chapter 3 you learned that climate change is a complex topic without straightforward solutions. Your students will ask bigger and more complex questions as a result of teaching about climate change that you may or may not have the answers to in the moment. Anticipate their questions and where they may face challenges with the content, and embrace those moments. We know that socioscientific issues (such as climate change) embody elements of Nature of Science because the data and information collected is on-going. You are not expected to know all there is about climate change because it has a tentative nature. New technology and scientific advancements allow for the collection of new data that will need to be studied. Learning more about how destructive human actions can be will push society to discover more innovative ways to solve the problem both at the individual and systemic levels. I am hopeful that future generations will one day read about how we were able to tackle this problem.

For your students, this is an opportunity to model what life-long learning looks like, to examine the true nature of science, and to provide them with the skills needed to explain what is happening in the world outside the classroom. You may not have the answers to their questions today, but you might be witnessing a future activist take interest in a problem they want to find a solution for in the near future. This chapter will guide you through facilitating and supporting your students in developing scientific literacy skills, critical thinking capacity, and thinking through how to create storylines using climate change as the anchor.

Using SSI to Develop Scientific Literacy Skills

What Is Scientific Literacy Anyway?

Scientific literacy has many different definitions that have evolved over time. The National Research Council (NRC) defines scientific literacy as, "[T]he knowledge and understanding of scientific concepts and processes required for personal deci-sion making, participation in civic and cultural affairs, and eco-nomic productivity. It also includes specific types of abilities" (1996). The NRC goes on to explain that a scientifically literate person is able to ask, find, or determine answers to questions that arise from everyday experiences. These individuals are able to describe, explain, and make predictions for natural phenomena. They are also able to evaluate the quality of scientific information and make claims based on credible sources of evidence. Lastly, these individuals are able to express positions on scientific and technological issues underlying national and local decisions as informed citizens. It is important to note that determining whether educators have been teaching students to be scientifically literate depends on their conception and definition of the idea. Anelli (2011) synthesizes different definitions of scientific literacy reflected in Table 4.1. Read through the definitions and determine the one closest to your personal beliefs about teaching and learning.

TABLE 4.1 Defining Scientific Literacy

Definition by Date	Description
Practical, Cultural, and Civic Scientific Literacy proposed by Shen (1975)	"Practical" refers to the application of scientific principles or technology needed to improve life. "Cultural" refers to the appreciation of science as a human achievement. "Civic" refers to the level of understanding needed to engage in science-related issues.
Civic Scientific Literacy proposed by Miller (1983, 1998)	Expounding on the previous definition, Miller defines civic scientific literacy as the level of understanding needed to read and comprehend science and technology related issues. They should be able to engage in societal debates that involve science and technology as informed citizens.
Functional Scientific Literacy	People need to have a foundational understanding of science because they cannot think critically about nothing. Even if most people will never hold careers in science, they will need to function as citizens and need to be scientifically literate to make informed decisions.
Scientific Literacy as Defined by NAS for the National Science Education Standards (1996)	"Scientific literacy is the knowledge and understanding of scientific concepts and processes required for personal decision making, participation in civic and cultural affairs, and economic productivity. It also includes specific types of abilities."
Fundamental and Derived Scientific Literacy proposed by Norris and Phillips (2003)	Fundamental scientific literacy has been simplified to only include reading, writing, accessing, and synthesizing information. These scholars argue that it should also include the ability to interpret, infer, analyze, critique, and contextualize information related to science. Students need to practice analyzing scientific texts with different intentions (an observation, causal relationship, generalizations, hypotheses, assumptions, supporting evidence, etc.) so they can also apply these skills to analyzing science in the media.

(continued)

TABLE 4.1 *(continued)*

Definition by Date	*Description*
Useful Scientific Literacy proposed by Feinstein (2011)	Science education needs to focus on the usefulness of scientific literacy. This includes the skills needed to solve personally meaningful, everyday problems, while also being able to make informed decisions on science-related issues. Feinstein argues that educators should teach students how to recognize science in "real-life" contexts where they can apply the scientific literacy skills.
Scientific Literacy in the NGSS (2014)	"One fundamental goal for K-12 science education is a scientifically literate person who can understand the nature of scientific knowledge. Indeed, the only consistent characteristic of scientific knowledge across the disciplines is that scientific knowledge itself is open to revision in light of new evidence."

Scientific Literacy in the NGSS

Recall that to support scientific literacy, the new framework embedded NOS principles within bundled performance expectations to support students engaging in three-dimensional learning. In part one of this book, you were introduced to the NOS elements and the complex process that scientists engage in. As we build students' scientific literacy skills and capacity, this requires teachers to be able to facilitate experiences that transition students from dependent to independent learners (with scaffolds that scale back over time). Research data affirms that when students are able to apply the NOS to complex socioscientific issues (such as the climate crisis), they increase their critical thinking capacity and scientific literacy skills through deep learning opportunities (Carter & Wiles, 2014; Khishfe & Lederman, 2006; Kolstø, 2001; Matkins & Bell, 2007; Sadler

 Which definition of scientific literacy aligns most to your beliefs about teaching and learning? Why might that be? How can you utilize climate change to develop students' scientific literacy skills according to the definition(s) you identified with?

et al., 2007). Consider engaging with Exhibit 4.1 to reconsider or affirm some of your own teaching practices that build critical thinking capacity.

Exhibit 4.1 Reframing to Build Students' Critical Thinking Capacity

In working with science teachers from various states, I have yet to meet two individuals who can describe NGSS-aligned teaching in the same way. These varying perspectives on NGSS have an influence on decisions made every day in the classroom. Below are statements commonly made by teachers that we can reframe into opportunities that build critical thinking capacity.

Statement 1 – "Lecturing Is Not NGSS-Aligned Teaching."

The word "lecture" itself does not allow for opportunities to co-construct knowledge because it implies that the teacher is talking *at* students during a teacher-centered lesson (think of this moment as the filling of a pail because students are answering basic recall questions as non-contributors). During a lecture, there is little to no productive struggle happening for students as passive listeners.

Reframing – Teachers can create a space where direct interactive instructional segments (not lecture) are those that include consistent opportunities for students to drive the lesson, and engage in sense-making to co-construct knowledge. First, think about how the lesson can be driven by students to engage in the complex process and nature of science. Consider the following questions as you facilitate your next lesson:

◆ Can you support students in posing sense-making questions that push them to revise or build on knowledge over time?
◆ Can you position students as the "drivers" of the lesson to build their capacity as contributors every day?

◆ Whose voice is often heard during these segments (from *both* teacher and student perspectives)? Whose voice is elevated or affirmed during these segments? Whose are left out of surface and/or deep level discourse?

◆ Can you facilitate discourse that would build students' capacity to do the heavy mental lifting to engage in sense-making (i.e. growing from productive struggles, believing in their capacity to engage, etc.)?

◆ How and when do you allow students to talk with each other during these segments?

◆ When are students able to explore their own interests regarding the central topic (allowing students to tap into their funds of knowledge)?

◆ What would be needed to make the above happen for you personally, logistically, and for all students (purposeful planning with vision of teaching in mind)?

Statement 2 – "Inquiry Labs Should Not Provide Students with Directions or Steps."

It is not uncommon for teachers to think that providing students with lab directions or scaffolded instructions equates to not being NGSS-aligned or inquiry driven. The National Science Education Standards (1996) defines inquiry as a wide range of practices students can engage with such as asking questions, constructing and testing explanations, communicating ideas, being critical thinkers, among others. You want to position students as capable doers of science and engineering, and you'll need to provide some initial scaffolding to support every student in the class to build capacity.

Reframing – Teachers can build students' critical thinking capacity around a central investigation that allows for them to have more choice and voice to engage with the process and nature of science. You can guide students through sense-making questions to help them access their cultural wealth on a topic they are equally invested in learning more about.

Perhaps we don't stress it enough, but we learn a great deal from our failed science experiments, which help us get closer to finding success. Similar to Statement 1, it depends on how you frame what students are trying to figure out, how they will figure it out, why they are engaging in this experiment to begin with, and what role you have during this lesson. Consider the following questions as you facilitate your next lesson:

♦ Can there be more than one approach to this experiment or activity? If not, how might you easily modify it to create space where students can drive the inquiry?

♦ Are students encouraged and/or supported in accessing their cultural funds of knowledge at any point? Is this intentional by lesson design?

♦ Why are students engaging in this lab or activity? What will the results be used for and how will students apply what they've learned? When is it explicitly linked back to the anchoring phenomenon?

♦ How is "failure" presented in your science class? Think about any subtle messages being sent about the process and nature of science when there is only one way to approach the experiment, one method to obtain the answer, one acceptable response, or one method to present findings/data.

♦ What sense-making questions are you posing to students throughout the experience? What responses do you anticipate?

♦ How will you facilitate their learning during this segment? What will you be doing, looking, or listening for?

♦ What are indicators of students engaging in the necessary productive struggle for sense-making compared to students that just need more guidance to build their confidence levels.

♦ Which NOS principles are they engaging with to build on or apply scientific literacy skills (i.e. employing a variety of methods, modifying claims in light of new evidence, communicating results of validity as a scientific community, etc.).

These are just two examples of common statements made when thinking about NGSS-aligned instruction. The goal is to continue to ask questions to examine your teaching practices as you encounter belief revealing statements. As you consider ways to anchor your instruction around climate change to empower students, continue to evaluate those lessons for scientific literacy and critical thinking capacity.

Using Climate Change to Develop Socioscientific Reasoning

As students begin to take more active roles in environmental justice issues that directly impact their lives, it is essential to develop their capacity to successfully lead changes in their communities. Rather than teaching climate change as isolated topics throughout the storyline, climate science can be used to ground the class as the larger issue students create solutions for. Approaching climate change through the SSI framework allows for students to engage in *socioscientific reasoning* that increases their content knowledge and understanding of the NOS for scientific literacy (Sadler et al., 2007). Socioscientific reasoning includes being able to recognize the complex nature of SSI, examine issues from different perspectives, understand the tentative nature of SSI, and exhibit skepticism when analyzing information (Sadler et al., 2007).

Sadler et al. (2004) uncovered that in order for students to understand NOS, students must first understand what constitutes data and its uses. Their research findings reveal that many students believe that the most convincing position is the one closely related to their beliefs. The second finding revealed that students were drawn to a position because it presents consequences directly related to them. As a result, researchers urge educators to challenge students by providing opportunities for reflection, discourse, and integration of scientific knowledge, while evaluating alternative views to align with scientific consensus (Sadler et al., 2004). Consider engaging students in the activities listed in Exhibit 4.2 to develop their socioscientific reasoning skills.

Exhibit 4.2 Opportunities for Socioscientific Reasoning

To further expose students to NOS principles, try implementing the following activities and lesson plans with students to build on their science and engineering practices. When teachers present more than one perspective to a problem, they increase students' critical thinking capacity by complicating their beliefs through rigorous instruction.

Two Sides Activity – Present the class with fictitious science briefings summarizing opposing positions. Half the class receives one briefing showing evidence by scientists who report on climate change as anthropogenic and a real threat, while the other half is presented evidence suggesting that climate change is a natural phenomenon and not at all an environmental threat. Have them gather data to support the claims made by the *briefings* (not their own claims leading to a personal debate), and then play the role of fact-finding detectives when they realize that both sides have "evidence" to support their scientists' claims (go to Bit.ly/FactDetectives). This activity also works well when presenting opposing positions on the benefits/dangers of Dihydrogen Monoxide (a.k.a. H_2O). Teachers can also ask students to sign an official petition to ban the substance before the big reveal (Bit.ly/DHMOban).

Global Oneness Project Video Clips and Lesson Plans – Have students learn about human impacts alongside climate science to frame your class discussions. When teachers have students engaging in argumentation over alternative sources of energy, basic human rights (i.e. access to clean air, food, water, etc.), or the degree of harm caused by climate change, students can engage in critical thinking by also considering stories told by real people impacted across the world. For example, when students consider the personal benefits of using fossil fuels, will they still feel that way when they see that it's at the expense of nature and other people/communities? Students can also participate in video and photography contests to tell their own stories at globalonenessproject.org.

NYTimes Climate Change Lesson Plan – Consider using the following lesson to teach about anthropogenic climate change through data analysis, impacts on people across the globe, and the complex nature of the climate crisis for critical discourse (visit Bit.ly/NYTLP). Help students to make sense of climate change data and to create an iterative perspective of the climate crisis as they learn more throughout your storylines.

Taking More Back to the Class

Now that you know the benefits of using climate change topics to support students' scientific literacy skills, it's time to take more back to class. Recall that the SSI framework stresses the importance of building instruction around a compelling issue and presenting it first to provide a true context for learning (Presley et al., 2013). These real-world contexts provide authentic experiences allowing students to have more depth of knowledge with skills to take action on issues outside the classroom (Hammond, 2015; Sadler, 2009). The following segments will help you learn more about phenomena, how to find strong ones, and how it can all circle back to the climate crisis.

Using Phenomena

A phenomenon is something that happens in the world that is not easily explained. There are two main types of phenomena (Anchoring and Investigative) you can include to help students make sense of what they are learning. Think of Anchoring Phenomena (AP) as big complex ideas, events, or questions that you can connect your everyday lessons back to. They are so complex that they involve many topics and concepts to fully grasp. The AP could take many weeks to fully understand and require several lessons, activities, lab experiences, discussions, etc. to fully unveil. The AP should also be open-ended and complex enough that every student can draw from a variety of resources to organize evidence needed to support their claims. Refer to Exhibit 4.3 for guidance on how to select strong AP for your storylines.

Exhibit 4.3 Finding Phenomena Criteria

Below are criteria to consider when you are selecting anchoring phenomena (AP):

1. A strong AP takes into consideration students, their communities, and their lived experiences. The anchor should be fascinating and meaningful to students in that they want to know more about what they are seeing or learning. Remember to think about the anchor serving as both "windows" into learning about and valuing other cultures, as well as "mirrors" where students see how they are represented and carry a presence in that lesson.

2. A strong AP requires students to engage in three-dimensional learning (DCI, CCC, SEP) to apply what they've learned to NGSS performance expectations. This process needs to be cyclical in that lessons intentionally connect back to the AP and revisited at different points across instructional segments. It should also be iterative in that the sequence builds from one day to the next and pushes students to engage in a necessary productive struggle.

3. A strong AP should be big enough for students to explore across several weeks because it requires more than one idea, concept, approach, or solution.
   a. Allow students to develop a class driving question from observing or learning about the AP. This allows

the claims made by individual students to be multi-faceted and dependent on the evidence they collect over several lessons to use in their explanations.

b. Make sure to let students know that evidence collected must be from class lessons, activities, labs, discussions, explorations, etc. to support their claim for explaining the anchoring phenomenon and driving question.

c. Deliberately call out the NOS or scientific literacy skills needed to fully construct explanations for their claims (e.g. students should be encouraged to modify their claims in light of new information or evidence over time).

4. A strong AP is observable. Students can observe the AP through video clips that don't reveal too much information, pictures, a class demonstration, a societal problem (i.e. invasive species, coral bleaching, looking at drought data, etc.), the context for your lab experiment, etc.

a. The AP should make students want to ask questions about what they are seeing. Furthermore, it should make them want to learn more about it to give science context.

b. Students should be engaging in the three-dimensions of NGSS across connected lessons in an iterative fashion.

5. A strong AP is important, relevant, and matters to students. It doesn't have to be phenomenal, but students should feel invested in learning more about what is happening to either create solutions for it, or take action when they have deep content knowledge on it.

As you think about which phenomena will anchor your storylines, note that using climate change topics fulfills all of the criteria above recommended by researchers and the NGSS.

Now that you know more about AP, let's explore supporting Investigative Phenomena (IP) and their role in storylines. IP are more direct ideas, events, or questions related (directly or indirectly) to the anchoring phenomenon. IP are used to provide additional information or clues to help unveil more about the main anchor. The answer to the IP could be uncovered the same day you introduce it, and you could have an IP for each lesson/lab so long as it clearly connects back to the anchoring phenomenon. It is crucial that teachers provide time for students to make sense of how their daily lessons connect back to the AP because this is how students can engage in sense-making. Figure 4.1 shows an example of how to start a storyline with an anchoring phenomenon, supporting investigative phenomena, and anticipated student driving questions.

NGSS Curriculum Development Storyline Tool

Collaborating and Designing with Students in Mind

As you begin to think about grounding curriculum on climate change topics for students, it will be tempting to use curricula developed by other schools or education companies. Although that might serve as a great starting point, your lessons need to be driven by your students and allow for co-construction of knowledge through iterative sense-making opportunities. Remember that our underlying beliefs about teaching and learning directly impact our teaching decisions. What might be unacceptable to us, may be the best thing since sliced bread for another teacher because of their differing underlying beliefs about teaching. Reflect on what you value as an educator as you create or comb through other people's lessons to ensure that it reflects big ideas of climate science, current research embedded in the framework on how students learn science effectively, and that it aligns to your vision of science education for every student.

We know that there will never be a perfect curriculum that works for every teacher. As you narrow your focus on creating or adapting resources, knowing your core values and beliefs

NGSS Curriculum Development Storyline Tool

Anchoring Phenomenon	Description: Students will watch a clip from Chasing Coral showing the New Caledonia Reef undergoing a bleaching event and causing the coral to fluoresce. *Credit Richard Vevers/The Ocean Agency	Connection to my students' interests, lived experiences, and funds of knowledge: Students will be intrigued by what they are seeing because it is highly unusual. They also generally care for living things and many of my students live near the ocean. As they discovered more about anthropogenic factors, they will become more interested in how they directly contribute to such events that are seemingly far away.
The Driving or Anchoring Class Question [Will be answered at the end and with evidence to explain the phenomenon]	I anticipate my class will generate one of the following questions: 1) What is causing the coral to change colors? 2) Why are the corals fluorescing? 3) What are the factors that lead to the corals producing the chemical sunscreen?	Learning this will help to address or solve a personal, family, or community issue by... I plan to use investigative phenomenon to teach students about their direct role and actions that lead to coral bleaching events. Also, I plan to show students how human health is reliant on the health of the ocean.
Potential Investigative Phenomena to Consider [Must explicitly connect to the AP and require students to collect data and evidence for]	Layered strategically throughout the storyline: 1) Watching a clip of faucet water catching on fire. 2) Watching a clip of river water on fire. 3) Looking at images of species that fell into Lake Natron in Tanzania. 4) Water and carbon cycle to understand how fossil fuels are created and role of limestone in the ocean. 5) Analyze data across the globe from NOAA on mass coral bleaching events. Ask more questions about the AP and possibly modify the AP based on data. 6) Reflect on human activities that would directly lead to an increase in CO_2 and change in pH in the ocean. 7) "Breath in a cup activity" 8) Acid-Base Titrations Group Inquiry Lab 9) Watching a short clip on ocean acidification to generate more questions. 10) Using science to regrow coral and help them become more resilient through music science.	Connections to the AP: 1) Students wonder about contaminants in water and how to filter it out. 2) Students learn more about what is in water and the role of how it impacts ecosystems and human health. 3) Students learn about potential of Hydrogen in water (pH), calculations for sense-making, and impacts on living things. 4) Students uncover more about human activities and CO_2 5) Students analyze larger sets of data to see patterns to show larger scaled impacts beyond the New Caledonia Reef. 6) Students connect their direct role to the AP and think about changes that need and can be made at different levels. Students also consider the complex nature of the topic regarding benefits 7) Students analyze and collect data to see how CO_2 changes the pH of water by breathing into a cup of water with indicator. They explore ocean acidification. 8) Students engage in an inquiry lab where they are provided with various unidentified acids with varying molar concentrations that they have to titrate and neutralize. This leads to learning about buffers and how the ocean acts as a buffer as it tries to regulate Earth's climate to reach equilibrium. 9) Understanding large-scale impacts on the ocean that impacts marine species and entire ecosystems. 10) Using science and engineering as tools to develop solutions.

FIGURE 4.1 Starting storylines with an anchoring phenomenon

about teaching and learning provides you with a stronger rationale and affirmation to make changes. To amplify the impact, it will be crucial to work with colleagues to determine the best curriculum for your students. Collaborating with colleagues should become easier because explaining why you want to advocate for certain components or ways of teaching to address equity issues (i.e. deliberate efforts to include discourse, start with meaningful phenomena, intentionally build in a culturally relevant lens, etc.) will be more clear. If a colleague disagrees with those lesson components or approaches, you will be able to listen with intent to help surface their values and beliefs about teaching and learning to have more productive conversations. When you're able to talk about the real problem preventing your team from moving forward (i.e. disconnect in teacher values and beliefs about how or what students should learn in science), you'll be able to have more effective solutions moving forward. As you begin creating storylines for your students, consider the following questions to guide the process:

1. What are the initiatives of your school (student talk, writing across the curriculum, focus learning targets, discourse, literacy, etc.)?
2. What are your science department planning priorities this year?
3. Looking at the NGSS state exam for your school, what areas of improvement were identified by the test?
4. What are your subject matter collaboration efforts like (if any)?
5. What is your vision of good science teaching? What might that look, sound, and feel like for students? What are potential common values shared among you and your subject matter teachers?
6. How would you describe the community you teach in?
7. What are the cultural backgrounds of your students? Do you know their deep cultural values (views of right and wrong, how knowledge is valued and passed down generationally, or ideas about individualism vs. collectivism)?

8. Have you had opportunities to get to hear more about your students' lived experiences? How are students affirmed?

When you have answered the questions above, consider engaging in Exhibit 4.4 to begin the process of creating new storylines catered to your students, school, and community.

Exhibit 4.4 Is This the Right Phenomenon?

 Think about a phenomenon that you are possibly interested in using for your lesson by answering the following questions:

- Is your phenomenon an anchoring or investigative phenomenon? Why?
- How excited are you about your chosen phenomena?
- What are the main science concepts students will need to learn to understand the anchoring phenomenon?
- Through what methods will they learn about those science concepts?
- Will they be exploring through activities and/or labs at any point?
- When and how will they learn NOS principles?
- Are there other phenomena you almost used for this unit that had more straightforward answers? Consider using these as investigative phenomena.
- Where can the investigative phenomena be placed throughout the storyline to continue

 Connect with the network to see how other teachers are integrating phenomenon-based instruction using at www.empowered scienceteachers.com (Book Resources → Chapter 4 → Discussion Board)

engaging your students and provide more information about the anchoring phenomenon?

Answering these questions will allow you to think about designing storylines that are directly relevant to your students, their communities, and their interests.

Finding Phenomena

Consider the following resources in Table 4.2 to start your search. It can be overwhelming and not at all helpful to search randomly, so remember to design with the end goals in mind. Which phenomena are complex enough to create storylines around, that also sequence and build on depth of content and skills needed over time? Also consider potential driving questions that you anticipate students developing through the selected phenomena to guide their learning journey. It might be helpful to test out different phenomena with different class periods to see which one produces stronger driving questions that can connect big ideas and topics. Given climate change's complex nature, anchoring instructional segments around the climate crisis allows for both cyclical, iterative, and relevant instruction to take place more fluidly. Looking ahead, Table 4.3 also lists several climate change phenomena currently used in science storylines to help inspire your design process.

Anchoring Instruction on Climate Change Phenomena

Climate change content makes for ideal phenomena because they are complex and require many different topics and ideas to fully understand. It is the issue that directly impacts students in every community, and an intersectional problem that calls for diverse solutions. As students are driven to take action on environmental and climate issues, we can support them by providing opportunities to learn about the complex nature of climate

TABLE 4.2 Phenomena Resources

Resources	Description
Access to Dr. Le's Padlet	Explore a variety of resources to integrate climate change into your curriculum. There are always updated current events that make for great phenomena. Bit.ly/LePadlet *Password:* GCCEducator
Explore Phenomena by Grade	Search by grade level and standards for general phenomena. thewonderofscience.com/phenomenal/
Phenomena for NGSS	Look at a compilation of video clips to use as phenomena. www.ngssphenomena.com/
Searching Phenomena by Topic	Type in a word search and by grade level to explore phenomena for your storylines. www.georgiascienceteacher.org/

change. If we want students to help come up with solutions for the climate crisis, grounding the content in climate change shows the urgency of this problem that unequally impacts everyone. Table 4.3 is a list of climate change related phenomena that you can utilize in your lessons and storylines. Consider accessing my Padlet for current events that also serve as engaging and relevant phenomena as well.

Connecting the Dots

Refer back to the curriculum map that you created in Chapter 3. As you think about integrating climate change phenomena into your units, you can add those AP or IP to relevant content topics (e.g. ocean acidification is an investigative phenomenon that I can add to my unit on acids and bases, the melting permafrost waking up ancient viruses to my segment on evolution, Earth's energy budget

 Consider submitting a Flipgrid video to the network for feedback or ideas. You can talk through your ideas, how you anticipate students will experience the phenomenon, and what you're looking to improve or gain ideas for at flipgrid.com/teachclimatechange (Password: Empowered).

TABLE 4.3 List of Climate Change Phenomena

Phenomena Ideas	Link (case sensitive)
Ancient viruses emerging from the permafrost	Bit.ly/PERMAfrost
Boiling seas (methane trapped in ice)	Bit.ly/BoilingSeas
COVID-19 Pandemic (Climate change and infectious diseases)	Bit.ly/covidngcc
Darvaza gas crater (release of methane gas since the 1950s)	Bit.ly/DARVAZA
Diver in Bali documenting plastic waste	Bit.ly/BaliDiver
Emerging "Hunger Stones" in Germany	Bit.ly/HungerStones
Extinction of a species in the wild (death of the last male northern white rhino).	Bit.ly/SudanPic
Faucet water that catches on fire (fracking)	Bit.ly/Wateronfire
Feedbacks – Looking at water runoff and snowmelt	Bit.ly/ioesSN
Florida "raining" iguanas	Bit.ly/IguanasFL
Ghost forests (impact on trees)	Bit.ly/GhostForest
Harvesting drinking water from thin air	Bit.ly/waterinair
Migrating spiders from Mexico entering California	Bit.ly/spidersew
MRSA or CRE outbreaks (antibiotic resistance)	Bit.ly/mrsaspread
Mysterious balls of goo washing ashore (stranded jellyfish)	Bit.ly/JFgoo
Ocean dead zones (agricultural runoff and human activities)	Bit.ly/AGdeadzone
Sea stars that rip their own limbs apart (wasting disease)	Bit.ly/SSspecies
Silicon Valley invasive species	Bit.ly/SVinvasive
Syberia's melting permafrost	Bit.ly/MPsyberia
The Sixth Mass Extinction	Bit.ly/6MExtinct
Traces of microplastic found in the most pristine places on earth (Arctic and Antarctic)	Bit.ly/Microplastic
Urban Heat Island Effect (self-reinforcing feedback loop)	Bit.ly/HeatEffect
World-wide coral bleaching events (start with Chasing Coral)	Bit.ly/WWbleaching

for understanding solar energy to learn about waves and electro-magnetic radiation, etc.). Once you feel that you have a handful of phenomena for that instructional segment, you can begin to piece your storyline together. This chapter introduces you to phenomena-based instruction with resources to help you focus your search. When you're ready to design whole storylines centered on climate change, Chapter 5 will provide you with teacher tools and resources to support the design process. Continue adding to your own curriculum map (see Appendix B) or refer to Appendix C for different curriculum maps created by science teachers thinking through big ideas grounded on the climate crisis.

Track your professional growth by re-examining your beliefs about teaching and learning.

Tracking Your Professional Growth

Student learning experiences in my class	To what extent do I agree or disagree with this statement. What are my current beliefs or values regarding the statement? Why?
Students often discuss policies related to science.	
Students have plenty of opportunities to collect and analyze scientific data or information.	
Students often discuss ethical issues related to science.	
Students engage with the Nature of Science principles.	
Students learn about and engage with the true Process of Science.	
Students co-construct knowledge with me every day.	
Students drive the instruction in my class as capable contributors and do-ers.	
Curriculum design elements I currently value	To what extent do I agree or disagree with this statement. What are my current beliefs or values regarding the statement? Why?
I currently build all lessons/units around anchoring or investigating phenomena.	
I present climate change issues at the start of each unit or lesson.	

Curriculum design elements I currently value	To what extent do I agree or disagree with this statement. What are my current beliefs or values regarding the statement? Why?
Students often engage in argumentation and making claims based on evidence.	
Students engage meaningful discourse opportunities every day.	
My lessons/units are centered around real-world issues that are directly related to my students lives or community.	
Students often use media/ technology to connect classroom content to the natural world.	

My current teacher attributes	To what extent do you agree or disagree with this statement. What are your current beliefs or values regarding the statement? Why?
I have to know everything about a particular issue before teaching about it.	
I know everything about climate science.	
I feel comfortable admitting to students when I don't know the answer to their question.	
I am comfortable teaching about open-ended issues where I cannot predict student responses.	
I have to feel like the expert in the room.	
I often experience imposter syndrome even for topics that I have strong expertise in.	
I have a strong understanding of the Nature of Science principles.	
I have a strong understanding of the NGSS framework.	

Refer back to the Book Introduction where you initially tracked your professional growth. Reflect on how your beliefs might have changed and why that might be. How might your responses unveil more about your teaching disposition?

Collective Voices for Climate Change Education

Emily Yam (Science Interpretation Manager at Aquarium of the Pacific)

In my experience, the most important thing to do is to link learners from the mechanism of climate change to **solutions** *that matter. People need to understand how they personally link to climate change, through the consumption of fossil fuels, and how climate change is impacting things they care about. When you understand how climate change happens, you can imagine how the challenge links to solutions that matter. These solutions are not just limited to consumer choices – climate adaptation matters, too. Things we do to lift our communities go a long way to make us more resilient to climate change. Everyone has influence with someone – even if it's just your family or friends. If fighting climate change is important to you, you can go one step farther by involving your family or friends. As an example: if you are in the position to try going vegetarian, invite family members and friends to join you, and you can try new recipes together. This creates connectivity between you and your family, and that connectivity is essential for climate resilient communities.*

Ruthie Gold (Education Consultant at Yale Program on Climate Change Communication)

One very important thing (it's so hard to only pick just one!) that teachers should teach students about climate change is that the majority of people in the U.S. believe that climate is happening and are worried about it! We often don't talk about climate change because we are scared that people will just shut us down, but more likely than not, the people you try to engage with will be eager to hear what you have to say. Be **brave** *and be* **vocal** *– the more we normalize talking about climate change, the better able we will be to affect positive change.*

Additional Teacher Resources

Access ESRI's resources to teach about nature of the environment –
Bit.ly/esriNATURE

Alternative Phenomena Criteria Checklist by Researchers –
Bit.ly/PhenomChecklist

Basic Physics of Climate Change –
Bit.ly/GCCPhysics

Check out essential questions for climate change by grade level –
Bit.ly/CCEssentialQ

Climate Change Teacher Resources –
CLEANet.org/

Harvard's Advanced Leadership Initiative on Climate Change Education –
Bit.ly/GCCHarvard

Listen to Leonardo DiCaprio's UN speech –
Bit.ly/BTFLEO

NYTimes resource to track temperatures in your hometown –
Bit.ly/hotterdata

Pinterest board by STEMeducation for indigenous science teaching resources –
Bit.ly/TRpinterest

Preview data in the U.S. impacting humans –
Bit.ly/Fractracker

Teaching Channel explains phenomena –
Bit.ly/TCPhenom

Use NOAA's Data in the Classroom –
https://dataintheclass room.noaa.gov/

Read ASCD's article on teaching scientific literacy –
Bit.ly/TeachingSL

Read the full 100kin10 2019 Trends Report –
Bit.ly/100k10Report

Watch clips from Conservation International series –
conservation.org/nature-is-speaking/

Watch "Kiss the Ground" to see how people are combating climate change –
https://kissthegroundmovie.com/

Want to know how scientifically literate you are? –
Bit.ly/MillerQuiz

What Americans know and don't know about science –
Bit.ly/PEWRESULTS

References

2019 TRENDS REPORT: *Trends and Predictions that will define STEM in 2020.* (n.d.). Retrieved May 13, 2020, from https://100kin10.org/news/2019-trends-report-trends-and-predictions-that-will-define-stem-in-2020

Anelli, C. (2011). Scientific literacy: What is it, are we teaching it, and does it matter. *American Entomologist*, 57(4), 235–244.

Caranto, B. F., & Pitpitunge, A. D. (2015). Students' knowledge on climate change: Implications on interdisciplinary learning. In *Biology*

Education and Research in a Changing Planet (pp. 21–30). Singapore: Springer.

Carter, B. E., & Wiles, J. R. (2014). Scientific consensus and social controversy: exploring relationships between students' conceptions of the nature of science, biological evolution, and global climate change. *Evolution: Education and Outreach*, 7(1), 6.

Hammond, Z. (2015). *Culturally responsive teaching and the brain: Promoting authentic engagement and rigor among culturally and linguistically diverse students*. Thousand Oaks, CA: Corwin.

Hansen, P. J. K. (2010). Knowledge about the greenhouse effect and the effects of the ozone layer among Norwegian pupils finishing compulsory education in 1989, 1993, and 2005 – What now?. *International Journal of Science Education*, 32(3), 397–419.

Hestness, E., McDonald, R. C., Breslyn, W., McGinnis, J. R., & Mouza, C. (2014). Science teacher professional development in climate change education informed by the next generation science standards. *Journal of Geoscience Education*, 62(3), 319–329.

Hodson, D. (2003). Time for action: Science education for an alternative future. *International Journal of Science Education*, 25(6), 645–670.

Kendi, I. X. (2019). *How to be an antiracist*. First Edition. New York: One World.

Khishfe, R., & Lederman, N. (2006). Teaching nature of science within a controversial topic: Integrated versus nonintegrated. *Journal of Research in Science Teaching*, 43(4), 395–418.

Kolstø, S. D. (2001). Scientific literacy for citizenship: Tools for dealing with the science dimension of controversial socioscientific issues. *Science Education*, 85(3), 291–310.

Matkins, J. J., & Bell, R. L. (2007). Awakening the scientist inside: Global climate change and the nature of science in an elementary science methods course. *Journal of Science Teacher Education*, 18(2), 137–163.

National Research Council (1996). *National Science Education Standards*. Washington, DC: The National Academies Press. https://doi.org/10.17226/4962.

Presley, M. L., Sickel, A. J., Muslu, N., Merle-Johnson, D., Witzig, S. B., Izci, K., & Sadler, T. D. (2013). A framework for socio-scientific issues based education. *Science Educator*, 22(1), 26.

Sadler, T. D. (2009). Situated learning in science education: socio-scientific issues as contexts for practice. *Studies in Science Education*, *45*(1), 1–42.

Sadler, T. D., Barab, S. A., & Scott, B. (2007). What do students gain by engaging in socioscientific inquiry?. *Research in Science Education*, 37(4), 371–391.

Sadler, T. D., Chambers, F. W., & Zeidler, D. L. (2004). Student conceptualizations of the nature of science in response to a socioscientific issue. *International Journal of Science Education*, 26(4), 387–409.

Somerville, R. C., & Hassol, S. J. (2011). The science of climate change. *Phys. Today*, 64(10), 48.

Part 3

Practices That Build Capacity for Student Agency

5

Planning and Teaching for Transformation

Read this when:

♦ *You're ready to move beyond phenomena-based instruction to designing meaningful storylines.*
♦ *You need guidance on NGSS-aligned and evidence-based teaching practices to utilize with these new storylines.*
♦ *You need concrete ways to support students to take action on climate change.*

Take a moment to learn about the following story which showcases innovation, persistence, and resilience in tackling a community energy and environmental waste issue. Carvey Ehren Maigue is an engineering student from the Philippines, who is currently developing renewable and sustainable energy using food waste. It started when he first noticed that the lens of his transition glasses would darken even on cloudy or rainy days. This led him to see how ultraviolet (UV) light is still able to reach him even when the sun's rays are blocked by clouds. Maigue decided to find novel ways to capture and convert this energy source. Examining his own community and cultural practices, he experimented with food waste as a potential game changer – specifically fruits and vegetables that contain organic luminescent compounds (they glow in the dark). Through numerous attempts, he discovered that these compounds are able to convert high energy UV waves into visible light. Using solar panels and films, he then turned this energy source into electricity. After many failed attempts and iterations, Maigue finally succeeded and his invention (Aurora Renewable Energy and UV Sequestration [AuREUS]) can now be used on buildings, homes, clothing, etc. to generate renewable electricity. Maigue went on to become the 2020 James Dyson Global Sustainability Award winner. We need to trust that people hold solutions to climate change in their own communities and remember that our students are capable of bold leadership. When we talk about student agency, we are referring to the capacity and ability to act that they already have within them. With guidance, modeling, and allyship, we can support students to take meaningful action on science issues that directly impact them.

A Vision for Science Education

When you think of a complex problem and find yourself overwhelmed on how to start tackling it, go back to planning with the end goal in mind (we know this as backwards planning). Essentially, you start by identifying a clear vision of teaching for your class (Could be the NGSS instructional segments, components of three-dimensional learning, climate literacy principles, etc.), and then you work backwards to unveil specific steps needed to achieve that vision. Being explicit with students about the learning process will only help them to better understand your methodology and get them to trust the process as well. I invite you to complete the activity in Exhibit 5.1 to see the larger picture when it comes to planning with climate change in mind.

Exhibit 5.1 Dream Bigger

Imagine that it is the year 2100, and humans have triumphantly solved the problem of curbing global climate change. Consider answering the following questions in this scenario.

♦ How does it feel?
♦ What will your neighborhood look like? Feel like? Sound like?
♦ What would people at the grocery store be doing? Buying? Using?
♦ Which species are thriving? Where do they live?
♦ What is our relationship with nature?
♦ What technology has been developed or advanced?
♦ Are there any changes to infrastructure, transportation, or public policy?
♦ What are students learning in science classes now for this new era?
♦ What government policies are in place to address anthropogenic carbon emissions?
♦ What changes need to happen and at what scale?
♦ What steps did we take to curb the climate crisis?
♦ What challenges did we address to bend the curve on climate change? Who achieved this? Whose responsibility was it?

As you consider your answers to the questions above, are there any steps you can take that get you closer to this vision? Regarding curriculum development, what can you include in your lessons that serve as indicators of success towards the goals you need to develop to achieve your vision of a healthier Earth? When you begin to find success with your students, how might you include your colleagues to amplify these efforts through shared leadership?

Backwards planning for gapless explanations is an approach often used to create NGSS storylines. Although previously referenced, this chapter will provide more depth with storylining for a shared understanding of the framework. Refer to the curriculum map you started in Chapter 3, and select one upcoming unit that you will teach. Then, access the storyline development tool online at www.empoweredscienceteachers. com → Book Resources. Write down any big ideas or concepts in the unit selected that you must address (set the parameters). Next, list smaller ideas and concepts needed to better understand the larger ideas or concepts you identified. Although you might have ideas on how to build out the storyline, consider the following questions:

1. Which phenomena are directly and indirectly connected to the key or sub concepts of this unit?
   ◆ Which are culturally relevant or responsive?
   ◆ Which allows for students to take action on issues they face?
2. What skills will students need to develop to learn the key or sub concepts identified?
   ◆ When can they build scientific literacy skills or develop socioscientific reasoning skills?
   ◆ How will they engage with NOS principles? How often? To what depth?
3. When do they engage in the process of science (doing science)?
   ◆ Is the problem open-ended enough to allow for students to tap into their own cultural wealth and funds of knowledge to begin looking for solutions?
   ◆ Who is driving the lessons and how are they positioned as a result? How strong is the alignment to the true process of how scientists, researchers, or engineers work?

This chapter will help organize your thoughts around these questions as you develop storylines grounded on climate change.

Overview of Storyline Design Components

Brian Reiser, a professor from Northwestern University who supports science teachers with NGSS, states that a storyline is a coherent sequence of lessons driven by students' questions that emerge from relevant phenomena. As you begin to revamp your curriculum, recall the importance of seeing the larger picture first then work backwards from that point. Exhibit 5.2 provides an overview of what storylines are and how lessons are sequenced.

Exhibit 5.2   Understanding Storylines

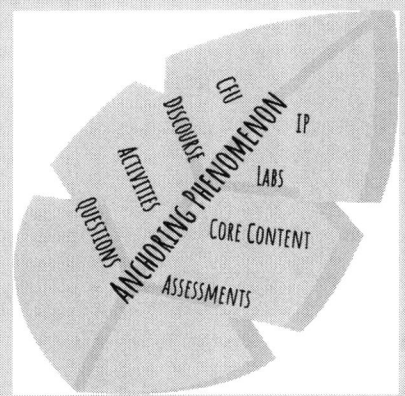

Imagine a storyline resembling parts of a tree leaf. When you look at a leaf, there is typically one central vein that starts from the bottom that may stretch all the way to the top. That central vein is your anchoring phenomenon (AP). When you look closer at this leaf, you start to notice smaller veins that deviate from the central vein. Those smaller side veins are the components of your lesson (such as investigative phenomena (IP), CER, discourse, lab experiences, activities, core content, etc.). What teachers tend to miss when designing the curriculum is how the smaller veins must always connect back to the central vein. In short, all that you teach in the unit needs to directly connect back to the anchoring phenomenon. Those lesson components should provide students with more information and evidence needed to explain the driving question developed from the AP.

As you build more storylines, imagine looking at different leaves that sit on the same branch. Although the branch itself represents the science course, the different leaves represent different instructional segments that are all connected through various anchoring and investigative phenomena. Can you imagine designing science curricula as a firmly grounded tree that branches off in different directions (subject areas), but always connects back to the climate crisis (the most pressing issue of our time)? Climate change is one of the only topics that allows for this to happen successfully because the problem is complex and requires an intersectional approach to learn and create solutions for.

With the bigger picture in mind, it's time to explore specific steps that will bring this curriculum to life. Figure 5.1 provides a visual to help you design storylines around an anchoring phenomenon (visit www.empoweredscienceteachers.com to access the online template).

Newmann, Smith, Allensworth, and Bryk (2001) argue that students are more likely to learn new content when experiences build on one another. The researchers also note that content learned through short-term exposure and referenced minimally will not be retained or transferred to other settings. When we think about how students learn, this does not come as a surprise to any skilled teacher. That is why the NGSS requires a cyclical approach to allow for students to revisit core concepts often as they build in complexity over time. If you look back at Figure 5.1, each lesson in the storyline relates to a larger driving question developed from the anchoring phenomenon to ensure that concepts are consistently cycling through the storyline. This iterative process allows for students to see and reflect on their own thinking as they gather data and evidence to explain the anchoring phenomenon.

Components of Strong Supporting Lessons

Let's address the daily lessons that are part of the storyline to ensure that you're not only teaching the words written in the NGSS, but that your pedagogical practices are also aligned to what the framework calls for. Recall that the NRC states that students learn when (1) teachers address their current understandings and cultural funds of knowledge, (2) they understand the true process and nature of science and are positioned as capable do-ers and knowers of science, and (3) they are given opportunities to be metacognitive for lessons

Storyline Unit Design Template

Instructional Segment		Course	
Semester		Weeks & # of Days Needed	

Setting the Stage: Storyline Overview

NGSS Performance Expectations:

Anchoring Phenomenon:

Reflect: Is it culturally relevant, meaningful (towards students or their community), allows my students to take action or apply science in some way, compelling, allows for students to engage in the SEPs to explore the topic's ethics, etc.?

Students already know and are capable of the following:

Anticipated Student Questions & Initial Driving Question:

Anticipated Student Questions -
Three questions I anticipate students posing from watching the clip of the AP are:
   1)
   2)
   3)

Initial Class Driving Question -

NGSS Three-Dimensional Teaching Components (DCI, CCC, SEP):

Other Standards Addressed:

NGSS Nature of Science:

CA Environmental Principles & Concepts (EP&Cs):

Climate Literacy Standards:

Common Core Math/English:

Anchoring Phenom	Investigative Phenom	Investigative Phenom	Investigative Phenom	Summative Assessment

FIGURE 5.1 Storyline Planning Tool

Lesson Checkpoints: Sequence & Flow			
Core Content and/or Anchoring or Investigative Phenomenon	What will students be doing?	Why are students doing this and how does it relate to the AP?	Where will students go next to build on their knowledge?
AP: Core Content Knowledge:			
Vocabulary words (based on core content of the day): -			
Student Self-Assessment - How will students know they are on target to move forward? What will they need to know?		1. 2. 3.	
Vocabulary words (based on core content of the day): -			
Student Self-Assessment - How will students know they are on target to move forward? What will they need to know?			
Vocabulary words (based on core content of the day): -			
Student Self-Assessment - How will students know they are on target to move forward? What will they need to know?			

Application of Knowledge: Performance Task
This could be an end of unit project, video, work of art, pre-post modeling, answering the driving question with evidence/data/information gathered each day and showing the modified claims, telling a story through pictures to apply what they learned (Taking action in the community in some way), inquiry lab experiment, etc.

FIGURE 5.1 *(continued)*

that serve as both "windows" and "mirrors" grounded in equity and antiracist teaching practices (2005, 2018). How might we teach science in ways that position students as the drivers of culturally relevant and responsive lessons? Like anything new, it will take time and practice to integrate all three elements into your daily lessons. More importantly, think about whether these research-based approaches align with your values as an educator (see Exhibit 5.3 for an example of what this might look and feel like). When you see every student engage in the productive struggle and build on their science identity, confidence, and agency over time, you will find yourself not wanting to teach any other way.

Exhibit 5.3 Lesson Design to Enhance Student Learning #1

The following is a breakdown of a lesson plan to show its skeletal components. This segment takes place on the first day of a new storyline.

1. Welcome students by going over focus learning targets for the day on the whiteboard (can be applying a science and engineering practice).
2. Introduce the climate change-related anchoring phenomenon using a video clip/picture/demo.
3. Facilitate class development of the driving question for the unit using students' personal questions or wonderings.
4. Make it clear to all students that the class's driving question serves as their summative essay question at the end of the unit in x weeks. This will help them choose a big enough question to gather evidence for, reflect on their learning each day to see how it connects to the AP, and be in control of their grade.
5. Groups of students share their favorite group question for consideration. The entire class votes on the one driving question they feel confident in being able to gather evidence for throughout the entire unit.
6. Students complete the "Question" and "Claim" portion of their Claim, Evidence, and Reasoning (CER) worksheet. They either write a claim or draw a picture of what they think might be happening for the phenomenon.
7. They share their initial thoughts and ideas with their elbow partner.

8. The teacher finally unveils the title of the instructional segment to start the lesson.
9. The teacher continues to introduce the first big idea with visuals while posing sense-making questions.
10. Students engage in small and whole group discourse every 5–7 minutes during the lesson where they pose sense-making questions to each other and try to figure out how the information relates to the driving question.
11. The teacher stops mid-lesson and tells students to talk with a partner about what they are learning so far and how it may serve as evidence to support, refute, or modify their initial claim for the driving question.
12. Students return to the CER worksheet to add evidence and re-evaluate their initial claim.
13. The teacher continues with the lesson by doing a short demonstration to showcase how the big idea works.
14. Students are prompted to return to the CER worksheet again to modify or add ideas or evidence.
15. The lesson closes by having students share with each other their current claims (which might have changed in light of new information), and evidence they gathered to support, refute, or modify their initial claim.
16. The teacher facilitates whole group discourse to hear students' ideas and understanding of content.

Although this is not the only way to begin a new storyline, it allows for you to see components of the lesson that address all three learning elements. Please identify which parts of the lesson (a) unveils students' current understandings, (b) allows for them to learn about the nature or process of science, and (c) pushes students to be metacognitive.

 *Consider answering the following questions to see where you are in your lesson planning, and what you consider to be essential in lesson plan design.*

- ◆ What was or was not surprising about the lesson plan above?
- ◆ Which learning element(s) do you already engage students in?
- ◆ Are students driving the lesson and learning often in your class?
- ◆ Are there learning elements that you wish to strengthen for students?
- ◆ How do you know what *all* students know?
- ◆ How do you know when students' thinking has grown or changed?
- ◆ How often or consistently do you allow for students to be metacognitive about their learning?
- ◆ Are they *all* engaging in discourse often enough? Writing practice?
- ◆ When do they employ critical thinking skills in the lesson?
- ◆ Now that you understand the learning elements, how might being intentional in your planning address student learning needs for NGSS?
- ◆ Which NGSS standards could be addressed in this generic lesson plan?

 *Looking at my answers and current teaching beliefs, my learning goals might be ...*

Student-Driven Instruction

Remember that the NGSS requires a cyclical and iterative approach. Although teachers may undergo various NGSS trainings, this central idea may never surface unless it is explicitly called out. Having a driving question allows teachers to revisit central ideas across lessons and instructional segments more easily. Giving students the opportunity to develop that driving question from

 These are components that you can start integrating into your lesson now. Take it back to the class and note how your class sounds, feels, and looks like when students engage in learning science with these elements. Which elements did you integrate, and which did you leave out? Why might that be?

the anchoring phenomenon is even more powerful. As many program teachers have shared with me, allowing students to generate the anchoring question for the storyline provides them with purpose for what they are learning. It also encourages students to make connections across lessons as they unveil evidence needed to make sense of the anchoring phenomenon. The level of engagement heightens for students that take ownership of their learning, and they become more invested. From the very beginning of the storyline, students are positioned as capable contributors whose ideas and funds of knowledge are driving the lesson. With this approach, teachers are able to provide three-dimensional learning for NGSS through a culturally relevant and responsive lens.

Launching Lessons with Phenomena

You can effectively facilitate the launch of the lesson with students as drivers in ways that are not "chaotic" or time consuming. Think back to the phenomenon you chose in Chapter 4 as the starting point. Another possible starting point could be the driving question(s) dictated by your school district curriculum map. Select a video clip, image, or demonstration you feel would help students to develop the driving question you already have in mind. To see if students will generate the driving question you are hoping for, practice by showing the phenomenon to non-science friends and ask them to provide you with three questions they are wondering about. If one of the questions provided was close to the driving question you had in mind, that is an early indicator that you selected a relevant video clip, image, or demonstration. Take the risk and put students in the driver seat to build their capacity as innovative young leaders. This small but meaningful act will allow teachers to gauge what students know, how much they know, what they want to know more of, and how far we can push them to reach their potential.

Approaches to Support Students Developing the Driving Questions

These are evidence-based approaches currently used by teachers with success to position students as capable contributors in the classroom.

First look. To introduce students to this approach, provide them with the driving question worksheet (see Appendix D) and post-it notes if possible. Show them a short (1–3 minute) clip or picture of a phenomenon without giving away too much context. Ask them to think about and write down one question they have about what they are seeing. Let the students know that they need to select one question as a group that if the teacher provided the complete answer for it, would explain everything about the phenomenon. After 60 seconds, assign small groups of three to four and have them all share their question. After giving each group two minutes to decide, go around the room and have each group share their chosen question out loud. During this time, support the entire class by having them write all the group questions being shared into the worksheet. Lastly, decide as a class by taking quick votes, on which question should be the class's driving question (see Figure 5.2 for examples). Remind students that the question chosen should be big enough to collect evidence for every day, and that they are deciding the essay question for their summative assessment. Troubleshoot issues that may arise by referring to Exhibit 5.4.

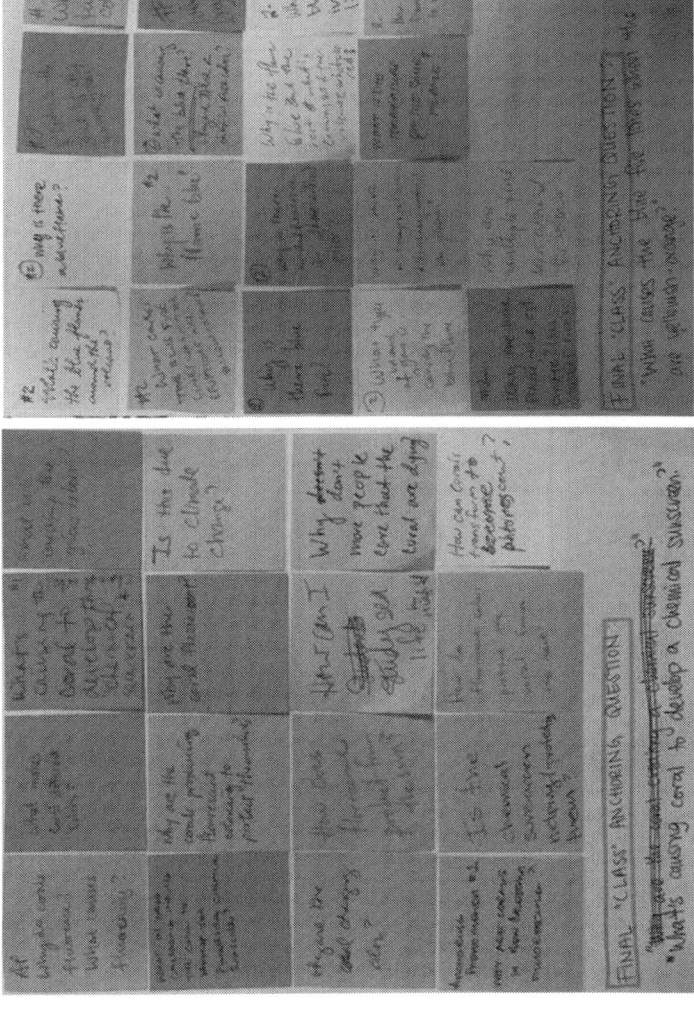

FIGURE 5.2 Generating Student Driving Question Samples

Exhibit 5.4 Tips for Supporting Student-Driven Questions

The following will help you to troubleshoot common occurrences with this teaching approach.

Question: What happens if different periods generate different questions?
Answer: Yes, that might happen! Over the years I have learned that each class varies slightly in wording, but generally they are all asking the same thing.

Question: How much information do you provide when you show the video clip, image, or do the demonstration?
Answer: I give students only basic information in the event they don't know what to focus on, or if they don't know what something is during viewing.

Question: Are you providing them with any other information during that time?
Answer: Every student must write one question down on the post-it note or worksheet during viewing in under 60 seconds. To support them as they are viewing, I typically say one of the following multiple times and circulate the room:"What is something you're currently wondering about?""What is a question that you have about what you're watching?"When I group them together to use their questions to develop the class' anchoring question, I lead them further by saying multiple times, "Don't forget, look at the questions from your group. You can only pick one to share with other groups so pick the one that if I answered it fully it would explain everything about the phenomenon."

Question: What happens if groups provide the same question when shared out loud?
Answer: I ask the group to share a different question presented to give the class a variety to choose from. I also don't tell students which question is the desired question to remain objective and allow for them to have more voice at the beginning of the unit.

Question: My class did not generate the driving question that I anticipated, what do I do?

Answer: Engage in the process of science as a teacher. So you didn't pick the right clip and their question wasn't exactly what you wanted, that tells you to either pick a different clip or adjust your facilitation of question generation with students (how are you guiding them without making it so obvious?). Make sure to have two different phenomena options to use in case the first one completely disappoints and reflect on your process for improvement.

Question: How long does this process (generating the driving question) normally take?

Answer: The very first time generally takes 25 minutes because you have to take time to explain what they are doing and what a phenomenon is. Beginning the second time, it will be a smooth 15 minutes before jumping right into the content.

Question: What if a student says they've seen the phenomenon and can look up the answer online?

Answer: Set ground rules at the beginning letting them know that they are investigators collecting evidence provided in class through lessons, labs, activities, discussion, etc. only. The point is to engage in the process of science and engage in argumentation using credible evidence provided from the class.

Second look. To continue engaging students in this iterative process, consider this second approach when students are comfortable for any AP or IP. Show students a picture of a phenomenon (like a picture of Sudan who was the last male Northern White Rhino, with his caretaker moments before the animal died and his species was pronounced extinct in 2018). Have students generate a question based on what they are wondering about, and repeat the process from the "First Look." After the driving question is determined, have students write down what they think the answer is for the phenomenon (this is their claim). Have them complete the CER worksheet and share their initial claim with an elbow partner. If appropriate, teachers can also use a mental modeling worksheet that allows for students to draw their initial ideas in one box. Have students add/modify their model every day given what they learned.

Third look. Once students begin to anticipate the type of question they should ask and select as a class (they developed this critical skill), challenge them by showing two clips of the same phenomenon that are slightly different to push them to modify their driving question to satisfy both events. For example, teachers can show a clip of a river on fire in Southwest Queensland to generate a class driving question, then show a second clip where a woman in the U.S. shows how the water from her sink catches on fire. Students are tasked to modify their question to ensure it addresses both phenomena prior to selecting a class driving question.

Fourth look. To get students to critically analyze questions, this next approach calls for them to go through the same process but using criteria to evaluate their group question prior to sharing it out loud. The criteria in Exhibit 5.5 can be provided to all students to help develop higher-level questions that call for more critical thinking. It can be used with any phenomena and with the same teaching approach.

Exhibit 5.5 Criteria for Developing Stronger Driving Questions

Students can use the following criteria as a checklist to strengthen their driving question.

◆ *It's Complex* – Is the question a yes or no question? If so, change it! What do you want to know that if I answered it would explain the entire phenomenon?

- ◆ *Definitely Relevant* – Does your question relate to something in the real world (man-made or natural event)? If yes, you're on the right track!
- ◆ *Open-Ended* – The question should not be too specific, understanding that there are many potentially answers. Some answers stronger than others depending on supporting evidence and credibility.
- ◆ *Take a Side* – The question should allow for you to make a claim and argue your position using credible sources of evidence from class.
- ◆ *Fascinating* – The question is interesting and seen as a mystery to you. You want to know more about this phenomenon through research, investigation, and/or discovery.
- ◆ *Scientific* – The question can be answered using the Process and Nature of Science.

Intentional Discourse Opportunities

Although providing discourse opportunities is not a novel idea, the following are tools to help make the process more consistent and meaningful for students. Recall the first lesson design component that fulfills both the NGSS and Student Learning Elements is unveiling students' current understandings and funds of knowledge. Context is needed to bring meaning to the content, and students need to make sense of what they are learning and why this lesson matters. As educators, if we're having difficulty seeing how the lesson relates to/or honors

 Consider submitting a Flipgrid video to the network for feedback or ideas on your storyline or lesson at flipgrid.com/teachclimatechange (Password: Empowered).

students' lived realities along with addressing 21st Century problems, it will be even more confusing for our students to know the connection. Are we teaching what students need to know to address the problems of today, or are we teaching science the way it's always been taught (same methods, perspective, and content that we learned)?

Consider answering the following questions to see where you are in your lesson planning, and what you consider to be essential in lesson design.

◆ What are your thoughts on the current state of science education? What are you happy with and what are you currently questioning?

◆ In what ways does climate change education allow educators to revamp science curriculum to be student-driven?

◆ How do you know what *all* students know during the lesson?

◆ What evidence indicates that students' thinking has grown or changed?

◆ How often or consistently do you allow for students to be metacognitive about their learning?

◆ Are they *all* engaging in discourse often enough? Guided writing practice?

◆ Specifically how are critical thinking skills taught and modeled for students in the lesson?

◆ What are the expectations that you have for students regarding discourse? Are these expectations explicitly shared with students with a provided rationale?

Looking at my answers and current teaching beliefs, my learning goals might be …

Developing Critical Thinkers in a Digital Age

Argumentation in Science Using Media

Often, teachers I support tell me that they typically use the CER tool during laboratory experiments. This is problematic because students are bombarded with claims through advertisements, social media, or conversations every day. Engaging in the process of science should be a daily occurrence for students in class, including learning to argue using credible sources of evidence and asking questions that evolve in light on new information. For example, analyze the information provided for the COVID-19 pandemic by different media sources. Table 5.1 shows the discrepancy in information that people were presented with during the pandemic. Please pause to review Table 5.1.

It is no wonder why there is a distrust in science given the way that information is communicated to the public. When teachers teach about the process of science, students understand the tentative nature of science and that claims can change in light of new evidence. When they engage in that process often in class, they will understand how science works without fostering distrust in the information provided by credible scientists. If we only explicitly call out the process of science during labs, students may believe that scientific reasoning skills do not apply in the real world. Students make claims every day for science or non-science related ideas. Being explicit about how to explore those ideas further using credible evidence, strong questioning abilities, and tools they can use to debunk misinformation is needed to ensure they become informed-decision makers.

Introducing Students to CER

Since argumentation is a skill that students need to become more informed-decision makers in society, teachers can introduce CER using media and technology to help students learn the general process. See Figure 5.3 for an example provided by a student showing their change in thinking over time, which resulted in changes to their driving question and initial claim.

TABLE 5.1 Protecting Yourself from COVID-19 Information

Information Released (First Month)	Information Released (Months 2+)
1. COVID can only survive on surfaces for a few hours.	1. COVID can survive on surfaces for a few hours depending on the material, but also for several days depending on other materials.
2. There is no need to wash your groceries.	2. You should take precautions with purchased goods because the virus can be transmitted through shared items.
3. The virus takes two weeks to have an impact on the body.	3. The virus can infect humans up to two weeks and people can start feeling symptoms as early as day 2.
4. COVID cannot infect your pets.	4. Several dogs have died from COVID.
5. Make sure that you go outdoors because the sun creates vitamin D for your body and the air is a natural disinfectant.	5. Stay at home to save lives.
6. COVID only targets the elderly with a small number of younger individuals (ages 20–40) with pre-existing conditions.	6. The U.S. death rates are showing that younger people without pre-existing conditions are dying from the virus (some weeks 40% of deaths were for this age group).
7. The U.S. death rates are continuing to rise making the country the new epicenter. Shelter in place orders issued to save more lives.	7. We have the virus outbreak under control with the current number of deaths at 100,000 (end of May), but under the current 200,000 early projections. Shelter in place order is no longer necessary.
8. Homemade masks made of cotton are ineffective at filtering viruses and nanosized particulate matter.	8. People cannot pick up essential items without masks to protect themselves or others from COVID. Masks are effective at catching viruses traveling in water droplets in the air. Wear a mask to save lives.

To give you a balcony view of the learning process, I am going to provide details of how teachers from the climate change education program learned to implement CER with students. It is important to position yourself at the level of your students to effectively anticipate student responses, what they know, what they notice, and what they might be curious about. Exhibit 5.6 is a breakdown of how other teachers learned about class facilitation of CER.

Claim, Evidence, Reasoning (CER) Worksheet

My Question:

My question: How do the white Rhino become extinct?

Driving Question: *Write additional questions you have as you learn more in a different color.*

What factors caused the white Rhino to go specifically extinct?
What factors cause species, like a white Rhino, to go extinct?

Claim: Why do you think this is happening?	White Rhinos are extinct because of human impact such as the people's needs for ivory and hunting.

Evidence: What science content, information, or data supports your claim? What would you use to convince someone that your claim should be adopted? This can be in bullet points.	• Biosphere includes all living things on Earth and is connected to all spheres
	 • invasive grass hazard near highway
	  – cars will light in middle of highway – grass will light
	  – Droughts = more fire b/c dried vegetation
	 • humans are at fault for intentional fire
	  – people demand = companies increase supply by destroying habitats
	   → destroying habitats = more pandemics due to mosquitos increase
	 • on mass extinction: what do we mean? who will it impact?
	 • humans are also species
	  – extinctions
	   sewerpipes
	   overfishing
	 5 extinctions (diff. causes but same ↑ CO_2)
	  • CO_2 — atmosphere
	    — ocean = ocean acidification
	
	 – top signs pointing to mass extinction
	 1) biological annihilation
	 2) species are threatened w/ extinction
	  → 1 million species
	 3) insects are declining
	  → bees = pollination = food source
	 4) amphibians decline
	 5) extinction domino effect
	  → ecosystem depend on each other
	 6) natural habitats are shrinking = species ↓
	 7) some animals have become so little = no role in ecosystem

Reasoning: How does the class information, data, or content support or make you want to adopt, modify, or reject your current claim (Write modified claims under your initial claim)? How does the data or information explain why the phenomenon is happening?	I would modify my claim to where it is more generalized to more species. Humans are heavily in the role of animal extinction. In various natural habitats, humans are creating intentional fires that destroys ecosystems. However, this only happens because regular consumers are increasing their demands to everyday packaged products. For example, chips and other packaged foods required palm oil in which companies set various areas on fire to create more land for supply. In different parts of the sea that used to have many sea life are now overrun by sewerpipes and overflowing to sustain ourselves. The extinction domino effect when one part of the ecosystem becomes extinct and disrupts the ecosystem causing other animals/organisms to go extinct. In addition, some animals, like koalas, are functionally extinct in which there are no longer apart in an ecosystem and will soon go extinct. (See page 2 for more evidence)

FIGURE 5.3 Iterative Student CER Worksheet Sample

CER SESSION 7
more evidence:
· 2 Billion people are estimated to become climate migrants by 2100
· Alicia Key - Let mein
   - WAR → affecting unarmed citizens
   - citizens fleed to Mexico → crossing the border
   - families separated
   - more than half Refugees are children
· heat index
   - heat + humidity
   - heat kills ppl
   - ↑ humidity + heat = ppl overheating (Deadly)
   - 125°F Chicago = 700 ppl died
   - 105°F + humidity = internal organs fail
· CArtools
   My address: 55-60%
      Pollution Burden 87   could be from
      PM2.5 66   freeways and constructions?
      Population 5.282
· Air conditioning brings more heat and makes the area hotter than it should be

I would modify my claim to be even more specific. Human actions can
negatively impact ecosystems of plants and animals by causing them to go extinct
But these actions also impacts ourselves. In various natural habitats, humans are
creating intentional fires that destroys ecosystems. However, this only happens because regular
consumers are increasing their demands to everyday packaged products.
In different parts of the sea that used to have many sea life are now overrun by Steel pipes and overfishing
to sustain ourselves. Human actions are heavily involved with the climate crisis. As from what we have
learned from the heat wave, our atmosphere is rising in temperature and mixing the heat w/ humidity,
our internal organs will fail and kill us. It has been estimated that 2 billion people will become climate migrants
in 2100. However, there are some parts of the world in which isn't open to the idea of opening their borders.
In addition, if countries do open borders, it will also effect the resources, animals and people currently
living in that place.

FIGURE 5.3 *(continued)*

Exhibit 5.6 Learning from Other Teachers' Experiences

*Background: The following program was provided between 2018
and 2020 across four non-profit environmental organizations to
secondary science teachers. Teachers from over 40 school districts
were represented and participated in 21 hours of professional
development on climate change education.*

Program Day Three – Introducing Teachers to CER

1. Using a CER worksheet, teachers were prompted to
   view a clip and write down the claim of the clip.

2. Teachers then watched a Super Bowl Kia commercial showing actress Melissa McCarthy driving around the world saving the planet in her new hybrid car.

3. Teachers then shared with each other what they believe the claim of the clip is. They were brought together as a group to share ideas out loud. Just as a person watching the advertisement at home, different people will generate different claims for what they are presented with.

4. Teachers were prompted to re-watch the clip again to collect evidence to support their individual claims. They were advised to write down anything they felt served as evidence to support their initial claim.

5. When the clip ended, they took turns sharing the evidence gathered to support their initial claim.

6. In order to help unveil teachers' rationale, they were asked to listen to each other and then adopt one claim that had the most credible evidence presented. The teacher will notice at this point that students begin to evaluate their own credibility of evidence.

7. Lastly, they engaged in the reasoning portion by evaluating the strength of evidence presented (if any). They had to decide to accept/modify/or reject their initial claim as a result.

Teacher Tip: Super Bowl commercials make for good introductions to argumentation in science using evidence. Popular social media events also work such as (1) The NASA Broomstick Challenge, (2) BBC Documenting Rare Flying Penguins, or (3) Liquid Mountaineering.

Teachers expressed concern about facilitating the "reasoning" part with students because they had difficulty doing so in the past. With CER, students will tend to reiterate what they already wrote as "evidence" to also serve as their "reasoning". I modeled for teachers how to unveil what students' current understandings might be in order to connect new information with what they believe is true. It is essential to unveil students' current understandings in order to address any misconceptions before introducing new content. Exhibit 5.7 has questions that teachers can pose during whole-class discourse to uncover students' reasoning on evidence selected.

Exhibit 5.7 Evaluating Credibility of Evidence Using CER

You can pose the following questions to help students think more critically about the evidence they selected to support their claim.

1. What do you know about the topic currently? How did you learn about that?
2. Why did you select that as evidence?
3. Is your evidence subjective?
4. Is it possible that someone in this class disagrees with your evidence?
5. Can you find credible and vetted data to support your evidence?
6. What is the source of your data?
7. Is what you selected actually considered evidence?
8. Would you classify your evidence as weak or strong? Why?
9. How does the evidence support your initial claim? If it doesn't, how can you modify your claim to reflect the evidence you have?

Imagine if your students firmly believed in specific ideas that weren't quite right for several years. As educators, we need to understand that it will take more than a couple of lessons to debunk these misconceptions. Worse yet, if the misconception is never unveiled then teachers operate under the assumption that students will just adopt new ideas and abandon their long-held beliefs. They might memorize the content for the test, but walk out with the same initial ideas they came in with. Transforming teaching means taking the time to know what students know so that they can have more depth of knowledge. Students must have opportunities to question what they know and how they came to know it to bridge the gap between old and new ideas.

There are several teaching resources listed in the *Additional Teacher Resources* section to help students strengthen their critical thinking skills as scientifically literate citizens. This section will highlight one of the resources by Tumblehome, but I encourage you to explore the additional resources listed at the end of the chapter as well. In order to provide more coherence between lessons in a storyline, ensure that each design element you integrate is purposeful and has clear connections to the anchoring phenomenon.

To help students resist scientific misinformation online, Tumblehome created teacher friendly lessons that are easily adaptable. Once students get comfortable with the consistent CER process in class, this resource supports students to use technological research skills to identify fake news. It is easily adaptable for the needs of your students and can serve as an introductory lesson on the importance of argumentation in science using evidence. Exhibit 5.8 is a list of topics that students can research using tools provided by Tumblehome as a good introduction to politically controversial climate change content.

Exhibit 5.8 Identifying Fake News Research Activity

Have students work in small groups to find out if they are victims of fake news. Using the "Resisting Scientific Information" teaching resources and tools, have them research their topic online. Before they look up the answers to their topic, have them engage in CER to generate a group question and a claim for their question. Once they are ready, allow them to research more about their topic online while writing down questions they may have. Lastly, have groups report out their findings and whether they accept/reject/ modify their initial claims in light of new evidence. Consider making this a regular class activity so that students can select current phenomena or current events to critically analyze that they are interested in sharing about.

Possible Topics (Can Be Current Events or Graphs to Interpret as Well):

1. Analyze the Heartland Institute Book for Teachers online denying climate change.
2. The green flash at sunset phenomenon.
3. Scientists have engineered artificial rain from clouds without any consequence.
4. Study showing all human feces have traces of microplastics.
5. 30,000 year old viruses are waking up as a result of the melting permafrost.
6. Sea stars are ripping their own arms off due to warming ocean waters.

Bringing It Back

As you ground your storylines around the climate crisis to empower students through information and action, take time to return to your teaching beliefs and values. Consider the following questions as you begin an important and necessary journey to disrupt science education. Think about why this work is important to you. Would anchoring your science instruction on climate change reflect your personal values and beliefs? What are teaching tools that you can readily take back to the class, and which ones are more challenging? Who benefits from this type of instruction? Looking ahead, Chapter 6 will push you to think more about the importance of addressing racial, social, and environmental inequities due to climate change. This crisis impacts every living thing on Earth, but some are more impacted than others. As young people rise up around the world to lead the climate crisis movement, teaching them about the complex and political nature of climate change will prepare them to be the climate warriors we need.

Collective Voices for Climate Change Education

Annie Frankel (Public Education Program at the California Coastal Commission)

When young people understand the basic mechanisms of climate science, they are **empowered** *to evaluate and advocate for solutions. Whether they can vote or not, they are capable of making an impact on the policy decisions of government at all levels, by expressing themselves in public hearings and to elected officials, and by sharing their knowledge with their family and community. You can't teach this subject without guiding your students to action. Effective climate change education necessitates civic engagement.*

Jessi Kerhner (Senior Scientist at EcoAdapt)

I think climate change **resonates** *most with people when it impacts (or is likely to impact) something they care about, such as their community, family, or a place they love, and when they feel like they can make a difference by something they can tangibly do. So I would focus on teaching examples of how climate change is likely to impact the things they care about as well as what adaptation (or combined mitigation-adaptation) options that students themselves could take (e.g., planting trees in urban areas to reduce the heat-island effect, advocating for/starting a petition to install solar panels on the school roof, engaging in citizen-science monitoring such as camera trapping of wildlife to look at migration patterns) in addition to those kinds of actions that decision-makers could take (and that students could potentially advocate for).*

Additional Teacher Resources

Consider how the NGSS DCI builds across grades and subjects as you select phenomena –
Bit.ly/DCIMATRIX
How the NGSS calls out SEPs by standard –
Bit.ly/ARGUING
Learn the scientific practices of explaining and argumentation –
Bit.ly/NGSSESA
Learn about using CER with The Wonder of Science –
Bit.ly/CERHELP
Review example storylines at Dr. Reiser's site –
Bit.ly/BRSTORYLINE
Sample checklist for phenomena with helpful tips –
Bit.ly/APIDEAS
Sample Claim, Evidence, Reasoning (CER) worksheet –
Bit.ly/CERWKS
Teaching resources to resist scientific misformation –
Bit.ly/TEACHRSM
Use the Bad Science Criteria poster with your students –
Bit.ly/BADSCIENCE

References

National Academies of Sciences, Engineering, and Medicine (2018). *How people learn II: Learners, contexts, and cultures*. Washington, DC: The National Academies Press. https://doi.org/10.17226/24783.

National Research Council (2005). *How students learn: Science in the classroom*. Washington, DC: The National Academies Press. https://doi.org/10.17226/11102.

Newmann, F. M., Smith, B., Allensworth, E., & Bryk, A. S. (2001). Instructional program coherence: What it is and why it should guide school improvement policy. *Educational Evaluation and Policy Analysis*, 23(4), 297–321.

6

Education for Climate Action

Read this when:

- ◆ *You want to support students communicating about climate change outside the classroom.*
- ◆ *You're ready to engage students in lessons that integrate climate, environmental, and social injustices.*
- ◆ *You're looking for ways to empower students to be agents of change in their communities.*

There is a real urgency and need to amplify our efforts as science educators to take on climate change for the sake of future generations. This work is challenging and time-consuming, but it can also be inspiring and serve as another way to excite students in STEM. Remember that students are bombarded with claims about the climate crisis every day, and rely heavily on teachers for scientific literacy skills needed to fight against disinformation campaigns and fake news. Thank you for being part of a larger movement to provide transformative learning experiences that will activate student agency. This final chapter will provide additional resources and tools to help you reflect on your teaching practices and ideas. Here you will find ways to help students get involved in taking action, which is just as important as teaching about the crisis itself. Rather than the "Gloom and Doom" approach that leaves students feeling disempowered, you can be intentional with opportunities to build their capacity as agents of change in their own communities. The socioscientific issues framework calls for teachers to address the ethical and social dimensions of climate change in order to help students access their agency as scientifically literate citizens. By providing them with a safe space to explore climate change, you are helping them to see science as a tool they can use to arm themselves against misinformation.

Tips on Talking about Climate Change

It is no surprise that students will experience a great deal of skepticism and push back when communicating about the climate crisis. This segment provides accessible entry points on how to talk about this politically controversial topic that aims to debunk common misconceptions. In general, it is essential to consider the following points when talking about climate change: (1) People need to know how climate change impacts them directly in order to care, (2) they also need to know how their daily actions connect directly to the impacts of climate change in order to reflect on their decisions, (3) the conversation must be hopeful and solutions based, and (4) they need actionable steps to take if they are thinking of making a change based on what they have learned. Complete the activity in Exhibit 6.1 to help students have successful conversations about climate change.

Exhibit 6.1 Listening with Intent Activity

 Students need practice to build their confidence in having conversations about climate change. Provide the following guidance for their consideration:

1. **Have positive presumptions** – Assume that the person you're talking to means well and they are not trying to attack you personally with their opinions.
2. **Listening with intent** – It turns out that people are generally not good listeners. If you find yourself trying to jump in constantly to talk then you're not listening. You're not listening while you talk. Let the person know you hear them even if they're wrong, so that way they give you the same courtesy when you speak about something they might disagree with.
3. **You are not there to change or convince this person** – Your job is not to change this person's values or beliefs through this conversation. If they are skeptical about climate change, chances are they have been for many years. Remember that one conversation with you is not going to change that, but multiple non-threatening conversations might.
4. **Ask questions for understanding** – One thing you can do is help them find holes in their logic, data, or reasoning. Ask questions about where they found this information and ask if you could look it up together. Ask about how they learned about that information and that you're just trying to understand a different perspective. Reaffirm that you're not there to change them.
5. **Plant seeds of doubt** – The thing about science is that it can be reproduced anywhere in the world by anyone enacting the same steps. When you hear misconceptions or incorrect data usage, don't jump at the person with data and facts. Look up that information together and take the stance that you are just trying to learn as well. This would be a good point to ask questions that get them to think more about the data.
6. **Engage in detective work** – What are their underlying values and beliefs about climate change? Is there a bigger reason for why this person may not want to accept the unrefuted data about climate science? A good scenario to pose is the following:

> a. *Say you went to the doctor and they ran your blood test at over 150,000 clinics and the results reveal that you have cancer. Would you believe the test results? What if they told you that they met with a panel of expert doctors treating cancer patients and they were 97% confident that you have lung cancer. Would you reject their diagnosis, refuse treatment, and pretend you don't have cancer?*
>
> This scenario essentially describes the scientific consensus around climate change that students should have already learned about at this point.
>
> 7. **Resources at the top of your head** – Make a mental note of the following resources as you look up information with the person you're speaking with. It's okay if you don't have all the answers, but you should know where to find answers from credible organizations.
>   a. Our Climate Our Future – ourclimateourfuture.org/
>   b. 5 Gyres – 5gyres.org/
>   c. Skeptical Science – skepticalscience.com/
>   d. Reach out to your science teacher for support.

Consider the following teaching resources to help students communicate about climate change successfully.

- ◆ Alliance for Climate Education – Teaching youth to have conversations with parents about climate change summary report (Bit.ly/ACETALK).
- ◆ Our Climate Our Future – Show students this video to help them talk about climate change with others (Bit.ly/OURCLIMATE).
- ◆ Bending the Curve Book – Chapter 3 provides information on how climate change directly impacts human health and disproportionately (Bit.ly/BTCBOOK).
- ◆ Watch Katharine Hayhoe's Ted Talk on communicating about climate change – (Bit.ly/CCTEDTALK)

Big Problems Require Big Solutions

Education either functions as an instrument which is used to facilitate integration of the younger generation into the logic of the present system and bring about **conformity**, *or it becomes the praxis of freedom, the means by which men and women deal critically and creatively with reality and discover how to participate in the* **transformation** *of their world.*

(Freire, 1972)

We must recognize that the time to disrupt science education is now if we are to protect the planet for future generations. As Paulo Freire points out, the educational system can be the vehicle to transform society if we recognize what isn't working and all the ways it fails to support every student in accessing their agency. Although not every student will commit to a STEM job or career path, the goal is ensure they are scientifically literate and have the option to pursue, succeed, and turn it down rather than feeling that it's out of reach. This requires that we deliberately move away from the old standards and acknowledge the huge paradigm shift in how science should be taught and reflected in schooling. It is incredibly hard work to dismantle an institution, but institutions are created by people and their beliefs can change.

It's currently the year 2021, and climate experts warn that we have less than ten years to bend the curve before we reach critical tipping points that will have devastating impacts for all life on earth (Intergovernmental Panel on Climate Change [IPCC], 2018). The future they warn us about is *here* as explosive wildfires ravage states, prolific hurricanes wipe away entire cities, ocean acidification is threatening the food supply of almost 1 billion people, the melting permafrost is causing sea-level rise and coastal erosion, deforestation is impacting whole ecosystems and unleashing harmful viruses and diseases, and so much more. Young people across the country are already mobilizing for the climate crisis, and we need to get behind them to propel them forward as climate warriors. Leveraging the research-based pedagogical approaches and climate science standards in the NGSS, take time to revisit what you believe is the purpose of science teaching is and what it has the potential to be.

Learning from Other Educational Leaders

In June 2020, CA State Superintendent Tony Thurmond, along with Dr. Pedro Noguera (Dean of The School of Education at USC), Sujie Shin (Deputy Executive Director of CCEECA), and Dr. Daryl Camp (President of CAAASA) gathered to talk about preparing students for the future amidst the pandemic. Their panel session focused mainly on shifting the paradigm for how educators teach underrepresented minority (URM) students in low-income communities and the growing inequities due to distance learning. Drawing upon culturally relevant and responsive teaching, Dr. Noguera noted that teachers need to recognize that it will take more than a week to learn about students' deep cultures, backgrounds, interests, and ways to effectively engage them.

We cannot ignore the fact that low-income communities made of predominantly URMs suffer more from the consequences of climate change, than affluent high-income communities that have the means to survive in increasingly challenging conditions. Although this panel session was not assembled address climate change directly, there were many points made that ran parallel such as (1) a paradigm shift is needed for educators to teach differently than what they personally experienced, (2) we need to teach in culturally relevant and responsive ways to engage students in issues that directly impact them, and (3) equity in the classroom is an intentional act by teachers that can disrupt traditional ways of teaching and learning. As an intersectional problem, climate change requires more than one solution and diverse ways of thinking. This complex issue requires all stakeholders to understand that an intersectional approach will bring about more meaningful, transformative, and systemic changes. (Please go to Bit.ly/TeachEquitably for the webinar recording.)

Climate and Social Justice

Those who contribute the least carbon emissions
will suffer the most and are currently burdened by this.
Don't wait for the political class to save you.
We need you to take action on this now.
(Former Governor Jerry Brown at ECCLPS 2019)[1]

We need courageous leadership for climate mobilization. As teacher leaders, we can elevate the voices of young people who are rising up to create solutions for climate and environmental injustices in their communities. It is essential to create change that people can feel especially when the injustices involve access to clean air, water, food, land – or education revealing that the lack of the former should not be accepted or normalized. For climate change, teaching through a social justice lens means that teachers embed opportunities for students to learn more about and make decisions on the ethical and political dimensions of the climate crisis. Remember that students will be bombarded with messages about climate change every day whether or not teachers decide to provide time to unpack these claims (Caranto & Pitpitunge, 2015; Carter & Wiles, 2014; Hansen, 2010; Hestness et al., 2014; Hodson, 2003; Matkins & Bell, 2007; Somerville & Hassol, 2011). If we choose to omit this part from our science curriculum (although the NGSS SEPs call for meaningful discourse opportunities to engage in argumentation from evidence, making claims, etc.), we are choosing the safety of silence on issues that impact students. We also miss the opportunity to prepare them to engage as young leaders who are both scientifically literate and informed-decision makers for society.

Move from Equity-Centered to Justice-Oriented

The first thing educators need to understand is that there are no strategies or a checklist for equity-center or justice-oriented teaching. It's about reflecting on your teaching practices, values, and beliefs and making intentional decisions to support every learner in your class. Consider asking yourself the following questions again to see if your thinking has changed or is further reaffirmed:

1. Think of a famous scientist. Who is this person? What is their cultural background? Do other famous scientists look like this person as well?
   Reimagined – Do the scientists that students learn about look like or identify with them? Is there value in having students learn about culturally relevant contributors in STEM fields as they develop their own STEM identities? Should we make an intentional effort to expose students to diverse scientists? Is there value to that?
2. Think of another famous scientist. How did you learn about this person? What does this person sound like? Do all scientists use the same language? Do all scientists express ideas in the same way?
   Reimagined – Who decides what scientists should sound like when they speak? How does our cultural upbringing shape our answers? When students express ideas in your class, do you notice *what* they are saying (content) or *how* they are saying it first (behavior)? Have you ever corrected the way a student expressed their ideas? What are you using as a guide when you have students rephrase or restate their response? Are there cultural ties to your view of who a scientist is, what they sound like, or their weight of their contributions?
3. Think of your science classroom. What takes up the most space on the walls of your room? Would an observer be able to identify your teaching values and beliefs in that space?
   Reimagined – In what ways might students see themselves represented in the classroom? Is that important as they develop their STEM identity? Are you consistent in making efforts to unveil and learn more about your students' deep culture? How might you elevate and affirm students' voices, ideas, and contributions so they feel valued in that space?

As you contemplate how science teaching could look, sound, and feel like with the questions above, Figure 6.1 will help you

FIGURE 6.1 The Social Justice Spectrum
Image Credit: Dr. Joseph Oteng @drjotengii – Social Justice Educator

to better understand the social justice spectrum to see where you currently are and where you want to be.

Incorporating Social Justice Issues into Your Storylines

As previously mentioned, those who contribute the least to global climate change, will be impacted the most. It is essential to dismantle oppressive and discriminatory systems that perpetuate the status quo. Recall that a powerful way to get students to think about climate change issues is to make it directly relevant to them. When teaching about energy, for example, teachers can tie in interactive tools (such as EPA TRI) to show where power plants are in the community as well as what chemicals are often released into the community soil, air, and water (and noting where it is not commonly found). Another example is when teaching students about aqueous solutions or properties of water, teachers can start with statistics (i.e. Water Crisis at Water.org) to show how nearly one billion people lack access to clean water each year or that every two seconds a child will die from a water-related disease (and how this basic human right will become a bigger challenge due to climate change). These examples high-light a systems-thinking approach to learning about science to reveal how interconnected issues are and direct human impacts.

As the keynote speaker at the 2018 CSTA, Dr. Ram Ramanathan strongly urged teachers to prepare students as climate warriors for this uphill battle to improve the planet they will unjustly inherit. He argues that he calls them climate warriors because, "this is a fight for survival to protect the habitat for generations yet to be born." Students need to be taught about the science of climate change so that they can develop solutions we desper-ately need to bend the curve. Students need to learn about the inequities of the world and in their communities, and they need to remain hopeful. It is essential that educators provide tools, resources, and support to channel students' desire to help better the world using science. Table 6.1 offers a few ideas that can be incorporated as anchoring or investigative phenomena that allow students to learn science through a climate and social justice lens (looking ahead, Table 6.2 lists organizations that build student agency for these issues).

United Nations Sustainable Development Goals (SDGs)

The United Nations SDGs were written to unite the world in addressing major goals that range from improving health, edu-cation, and inequity, to taking on the climate crisis and miti-gating its impacts. Teachers can use the SDGs to center curricula around real-world issues and connect with educators around the globe working towards the same goals. Figure 6.2 identifies all 17 sustainable development goals that teachers can explore more in depth and find teaching resources for.

TABLE 6.1 Climate and Social Justice Issues with Phenomena

Climate and Social Justice Issues	Possible Phenomenon
Air Pollution and Impact on Specific Communities	Use this resource to analyze air pollution by looking at the "pollution burden" score and health scores by zip code. Generate student questions, and find ways to move to civic community engagement based on the revealed data. *Resource: Bit.ly/CAtool*
Clean Water and Impact on Specific Communities	Use these resources to reveal contaminants in water by zip code. Are there ways to protect communities from these contaminants? In what ways might education on water quality lead to community action based on the data? *Resource: Bit.ly/CAtool or www.fractracker.org/map/us/*
Environmental Justice and City Responses	Explore your city's Climate Action Plan (CAP) to see how they define and address environmental justice implications of climate change. *Resource:* www.ca-ilg.org/climate-action-plans *Specific Example:* Los Angeles City pLAn (See Chapter 1 Environmental Justice) https://plan.lamayor.org/
Fracking and Impact on Specific Communities	Use this resource that maps out each state and the influence that fracking has on water, land, and air quality in each community. What data can be drawn and analyzed from this site? How might education on fracking protect communities in the future? *Resource: www.fractracker.org/map/us/*
Heat Island Effect or Sea Level Rise Impact on Communities	Use this resource to analyze high heat days by zip code to show the heat island effect or sea level rise in different communities. What will happen when there are power outages as the heat index increases and demands more energy? How are climate models showing sea-level rise and collapse of ecosystems in your area? How might education on sea-level rise protect communities in the future and lead to better city planning? Use this to spark student conversations and move to community action. *Resource: https://cal-adapt.org/*
Hospital Fatality Rates and Impact on Specific Communities	Using information provided on hospital malpractices to determine hospital grades, what patterns can be seen in lower rated hospitals compare to highly rated ones? What can we do with this information and move towards community action? *Resource: Bit.ly/HOSPITALGRADES*
Housing Inequities and Impact on Specific Communities	Using Images of low-SES and high-SES communities to show disparities, take questions, explore implications, and provide discourse opportunities. *Resources: Bit.ly/PHOTOINEQUALITY* *Specific Neighborhood Example Resource:* UCLA's "Beyond the Schoolhouse" report for Los Angeles Bit.ly/BeyondTheSchool.
Wildfire and Air Quality Data Tracker	Use this resource to analyze recent changes in air quality by zip code. Have students analyze the graphs and anomalies they identify. Have them zoom out to look for wildfires nearby and data trends. *Resource: https://fire.airnow.gov/*

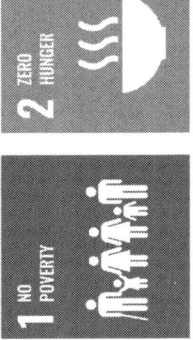

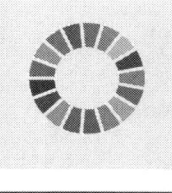

FIGURE 6.2 United Nations Sustainable Development Goals

Image Credit: www.un.org/sustainabledevelopment (The content of this publication has not been approved by the United Nations and does not reflect the views of the United Nations or its officials or Member States)

Resources That Build Student Capacity

> Science is constantly proved over time. If we take something like any fiction and destroy it, in a 1000 years time they would not come back just as they were. If we took every science book or every fact and destroyed them all, in a 1000 years they would be back because all the same tests would have the same results.
>
> (Ricky Gervais)[2]

When Ricky Gervais was being interviewed by late night show host Stephen Colbert, he confidently defended his stance about trusting science on national television with the quote above. We need to arm our students with resources, tools, and approaches they can use to respond confidently when confronted by a climate change skeptic. It is clear that this complex problem requires for people to work together to actively find solutions. This segment will provide teachers with pedagogical approaches to support students through co-constructing knowledge together.

Dictogloss Activity

Dictogloss is a classroom activity where students learn to use one another as resources of information to reconstruct a short text. This activity positions all students as capable learners and contributors of information through collaborative efforts. Students also learn quickly that they need each other in order to complete the task because everyone can contribute. The following is an overview of how to facilitate the activity with your students.

1. Start with a current event, scientific article, or science concept with multiple parts. Let the students first analyze on their own to find big ideas, identify questions they have, and try to make sense of it. Then provide time so they can share through discourse their level of understanding with an elbow partner.

2. Next, go over the ideas as a class and let the students know they get one chance to write down or draw as many details as possible.

3. Have students share what they were able to write or draw with each other and their level of understanding.

4. Let them know that you'll be going over it one last time and their job is to piece together the current event, article, or science concept with as many details as possible. This time, tell them they can work in groups of four to come up with a plan for the second listen.

5. After the last exposure, allow for them to work together to reconstruct the current event, article, or science concept with as much details as possible and to indicate what issues they might have encountered preventing them from fully understanding the reading.

6. Facilitate a discussion about how this activity emphasizes the need to work together and to use one another as resources of information. Emphasize how this positions them as scientists because scientists rarely work in isolation and collaborate in teams to help see the full scope of the problem to fully understand the problem.

More Science Teacher Talk Moves

Consider using talk moves for productive conversations in your class (refer back to Chapter 1, Exhibit 1.8). Whether you are facilitating whole or small group discourse, think about how equitable those experiences are for students. Do you find that students tend to engage in dialogue with you as their teacher, instead of their peers? When they share their thoughts and ideas during whole group discourse, are they confident in participating and feeling open to having their ideas challenged? Consistently using talk moves allows for teachers to support students to talk with each other to

Take it back to the class with this Dictogloss worksheet to help facilitate this learning segment (Bit.ly/ DictoglossWKST).

help them engage in the process of science as a community of learners. Scientists communicate with one another often and they also have to think about how they will share their research in accessible ways. Build up your students' abilities with the following resources:

1. STEM Teaching Tools Talk Flowchart – stemteachingtools.org/brief/35
2. Talk Moves Goals Reference Sheet – Bit.ly/TALKMOVES
3. Exploratorium Science Talk Tool – Bit.ly/ExploratoriumST

Exploring "I Don't Know"

On the Cult of Pedagogy, Connie Hamilton shared a useful tool during her interview to help teachers to explore what students mean when they say, "I don't know (IDK)." Turns out that when students respond with, "IDK" there could be a variety of reasons that move beyond not knowing the answer to your question. Are they saying, "IDK" because they don't understand the question? How helpful would it be for a teacher to hear the student say instead, "I didn't understand the question, can you rephrase it for me?" How might the teacher respond knowing the exact point of confusion rather than dealing with "IDK" broadly? Hamilton also notes how powerful it can be to just add the word "yet" to their response to let them know that they are still growing and that it's okay to not know yet. She states that when students don't know something, we need to transition them to knowing more about that thing, rather than terminate that line of thought. I recommend downloading the resource on the website and providing a copy to each student to utilize when they are engaging in discourse. Simply asking if they can have more time to think about the question because they need more processing time, reveals so much more to you as a teacher then IDK.

 Take it back to the class with this resource to help facilitate this learning segment Bit.ly/HandlingIDK.

Gauging Student Progress

Mini-Lab Portfolios

Creating mini-lab portfolios allow you to enhance the traditional lab reports and position students more authentically as scientists. Students can complete this using an online format (Using Google Slides, Keynote, Prezi, etc.) or a physical mini-presentation board using recycled manilla folders. Remember to scaffold instruction on writing these segments so students are not overwhelmed and are working with supports. It is generally a good idea to have the first portfolio completed in lab groups of three to four, then the second portfolio in pairs, and eventually completed individually.

 See Appendix F for example student portfolios and the template to use with your students. Access these online at www. empoweredscienceteachers. com (Book Resources → Google Drive)

Students as Experts

Once your students have gotten used to creating mini-lab portfolios, it's time to position them as scientists by sharing what they have learned with the community. Have students present their portfolio to one adult on campus in 5–7 minutes. Allowing students to teach their history, math, or art teacher about science builds their confidence because they are the experts in that moment. Have the adult listener submit a quick Google Survey evaluating the students' presentation so you can provide students with feedback afterwards. Normally one week is enough time to complete this activity. Encourage students to look at the Google Survey prior to presenting so they can anticipate the questions. It is helpful to send out an email to your administrator to let them know that students will be reaching out to adults on campus so that all teachers know what to expect.

 Access these resources online at www. empoweredscienceteachers. com (Book Resources → Google Drive)

Science Oral Defense

Gauging students' level of understanding can also be completed through an oral defense at the end of your instructional segment. This is an alternative method of assessing students by allowing them to present their knowledge through discourse with their peers. Students are provided guidelines prior to the defense so they can prepare their 10-minute segment. In groups of three to four, one student will present their big ideas with supporting evidence while the other students listen, take notes, and evaluate the strength of information/evidence provided. The grade is determined by the average of all the listeners and your final evaluation as you circulate and listen. Allowing students to explain what they feel are the big ideas and what serves as evidence is another way to support different learners in the class while engaging in the Process and Nature of Science.

 Access these resources online at www. empoweredscienceteachers. com (Book Resources → Google Drive)

STEM Teaching Tools

If you are looking for ways to better align your assessments to gauge student learning on a deeper level, consider exploring these resources from the *STEM Teaching Tools*. This database has online professional development and a variety of resources to integrate three-dimensional learning. For grades 6–12, the first resource allows you to integrate the Cross Cutting Concepts (Bit.ly/CCCQUESTIONS), and the second resource allows you to integrate Science and Engineering Practices (Bit.ly/SEPQUESTIONS). These resources help to modify current questions for assessments while providing pocket questions to pose for class discussion around any topic.

 Take it back to the class with this resource to help facilitate this learning segment at http://stemteachingtools.org/.

Interactive Tools to Enhance Instruction

Climate change content needs to be directly relevant to students' lives to show the importance of why humans must take action now to mitigate the impacts and bend the curve. These interactive tools allow teachers to use data to highlight the impacts on students' communities (many by zip code). Position students as the drivers of instruction by having them use, interpret, discuss, and pose questions they have about the data. Students can analyze by city, state, or country, which allows them to better understand the larger societal and systemic inequities or injustices using data. Recall that 100kin10 predicted that environmental and climate justice issues will move to the forefront of education because of student demand; these tools provide the historical context needed to better understand the issues they care most about. Table 6.2 outlines top interactive tools that teachers can use to engage students in science practices to see direct community impact.

Empowering Students to Take Action

Remember that climate science should not be taught in isolation, represented over a few disconnected lessons, or taught without opportunities to explore possible solutions. We cannot engage students in science practices when lessons are not grounded in real and culturally relevant phenomena. The following segment provides teachers with resources on supporting students to take action in their own communities as capable scientists or engineers. Consider taking risks to provide intentional opportunities for students to explore the ethical dimensions of climate change to also support them as young community leaders. Lastly, move away from instruction that disempowers students (i.e. positioning them as only receivers of knowledge with no opportunities to contribute, doesn't affirm or elevate the cultural wealth they possess, or doesn't provide space for them to engage in the process of science in ways that are personally meaningful).

TABLE 6.2 Reputable Interactive Educational Tools

Online Tool and Access	Description and Uses
CalEnviro 3.0 **OEHHA –** California Office of Environmental Health Hazard Assessment *Access:* Bit.ly/CAtool	Based on the census tract, students can look up their neighborhood's population burden score. The higher the overall number, the higher the burden for that population. It also measures toxicity in the air, water, and soil as well as social and demographic information by zip code.
Climate Interactive – Free tools to simulate/model solutions for climate change Access: Bit.ly/CCINTERACTIVE	Have students engage with models to analyze and interpret data for NGSS and to engage in science practices. These models will inform students' understanding through their programs: En-ROADS, C-ROADS, and ALPS.
EPA TRI – U.S. Environmental Protection Agency Toxic Release Inventory *Access:* Bit.ly/EPATOOL	This online interactive tool allows students to look up what toxic chemicals are released in their neighborhood by zip code. They can see what chemicals are released into the air, water, and soil as well as facilities in your neighborhood.
Fire and Smoke Map – Interactive map by zip code *Access:* fire.airnow.gov/	This online interactive tool lets students analyze fires by state and zip code to analyze air pollution quality within the last week. Type in the zip code, hover over the censor (dot), ask questions, zoom out for larger analysis and to generate more questions.
NOAA Data in the Classroom – Interactive Tools *Access:* dataintheclassroom.noaa.gov/	Explore a variety of interactive tools to enhance student learning in the areas of coral bleaching, sea level, ocean acidification, El Nino, and Water Quality. Each tool comes with teacher lesson plans and student worksheets.
Population Earth – Teacher Resources *Access:* Bit.ly/POPEDUCATION	Explore this resource to teach about the current population numbers and use data to introduce concepts like carrying capacity, shortage of resources, human impact on Earth, etc.
Prezi Climate Change – Teaching Lessons *Access:* Bit.ly/PreziCC	Explore a variety of lesson slides on climate change content, and their video platform that allows you to create virtual lessons as well.
The New York Times – What's Going on in This Graph? *Access:* Bit.ly/NYTGRAPHS	Experts in a variety of fields (including climate change) post graphs once a week to engage students across America in analyzing and interpreting data.

Community Science Experiences

You can have students support research efforts in a variety of ways by integrating community science experiences in your curriculum. These programs allow students to collect or analyze data for researchers in their community. The data is often used to inform decision-makers and students can be part of that experience to learn more about the process of scientific research.

 Take it back to the class and check out Dr. Le's second Padlet dedicated to community science programs at Bit.ly/CommunitySci.

These programs are also great ways to position students as capable scientists where they can be experts in the process and share what they learn with those around them. Explore Table 6.3 to connect with organizations across the nation that build students' agency.

TABLE 6.3 Organizations That Build Students' Agency

Online Tool and Access	Description and Uses
5 Gyres – Students can be community ambassadors. *Access:* www.5gyres.org/ambassadors	Students will have access to resources to host community events as an ambassador and ways to connect with others across state lines. There are also research studies that take place in different states every year to help lawmakers make informed decisions using data collected by community/citizen scientists.
350 – Connect with like-minded people internationally and get involved in your own community. *Access:* https://350.org/get-involved/	Join the international movement of people trying to end fossil fuels for a sustainable world. *"We believe in a safe climate and a better future – a just, prosperous, and equitable world built with the power of ordinary people. - 350"*
Climate Generation – Connect with students across the nation taking action on climate change. *Access:* www.climategen.org/	Teachers can access free curriculum resources and information on professional development opportunities. Students are also able to act on climate change by joining the Youth Environmental Activists network.
Ocean Conservancy – Be an ocean advocate and take action. *Access:* https://oceanconservancy.org/action-center/	Students can become involved as advocates of the ocean by taking action on marine life, protecting the Arctic, climate change, and the health of the ocean.
Our Climate Our Future – Promote student activism and networking. *Access:* https://ourclimateourfuture.org/	Get students involved in community activism and help them to connect with like-minded students across the nation on the topic of climate change. Teachers also have free access to teaching resources.

Online Tool and Access	Description and Uses
Strategic Energy Innovations (SEI) – Teacher curriculum and student certification programs for future green careers. *Access:* www.seiinc.org/	Explore free NGSS aligned science teacher curriculum on topics such as energy, climate change, and biomimicry. Students are also able to earn certification for building solar cells, completing an energy audit, and others to build up their resumes for future green jobs.
Sunrise Movement – Connect with like-minded people internationally and get involved in your own community. *Access:* www.sunrisemovement.org	Join the movement of young people creating a greener future and bending the curve on climate change. *"Together, we will change this country and this world, sure as the sun rises each morning. – Sunrise Movement"*
The Biomimicry Institute – Have students engage in the Youth Design Challenge. *Access:* https://biomimicry.org/	TBI prides itself on supporting students to develop skills, be curious, and gain new perspectives through their educational resources. *"Biomimicry is a practice that learns from and mimics the strategies found in nature to solve human design challenges – and find hope along the way. – The Biomimicry Institute"*
The Years Project – Everyone can take action in various ways with TYP. *Access:* https://theyearsproject.com/	Explore accessible ways to take action in your community and connect with fellow teachers or students. They also have teaching resources (video stories, clips, lessons, etc.) for teachers through their climate classroom portal.

Concluding Remarks

Thank you for your time, dedication, and commitment as an educator and an influential agent of change. We have a huge responsibility to teach about the science of climate change in ways that are sobering, hopeful, and empowering. As young leaders across the world mobilize to demand action on climate change, we can build their skills as informed decision-makers and solution designers with science. We need to prioritize consistent and culturally relevant opportunities that allow students to take ownership of their own learning experiences in our daily lessons. Have confidence that this will push them to access their creativity, passion, innovation, and resilience to bend the curve.

As you move forward with anchoring your instruction around the climate crisis, remember that you are not alone in bringing climate change through a social justice to the forefront of education. Connect with the online community of educators to either give or receive support. Reach out to your teaching colleagues to see who is ready to take on the challenge. Explore local and national organizations that are doing something about the climate crisis. Intentionally draw upon teaching practices and resources that build capacity in your students as young leaders and community change agents. Finally, recognize the rippling impact that you make as an educational leader. We are supporting the next generation of climate warriors, and we *will* succeed because we will do this *together*.

> *The new dawn blooms as we free it.*
> *For there is always light,*
> *if only we are brave enough to see it –*
> **If only we are brave enough to be it.**
> (Amanda Gorman)

Track your professional growth by re-examining your beliefs about teaching and learning.

Tracking Your Professional Growth	
Student learning experiences in my class	*To what extent do I agree or disagree with this statement. What are my current beliefs or values regarding the statement? Why?*
Students often discuss policies related to science.	
Students have plenty of opportunities to collect and analyze scientific data or information.	
Students often discuss ethical issues related to science.	
Students engage with the Nature of Science principles.	
Students learn about and engage with the true Process of Science.	
Students co-construct knowledge with me every day.	
Students drive the instruction in my class as capable contributors and do-ers.	

Curriculum design elements I currently value	To what extent do I agree or disagree with this statement. What are my current beliefs or values regarding the statement? Why?
I currently build all lessons/ units around anchoring or investigating phenomena.	
I present climate change issues at the start of each unit or lesson.	
Students often engage in argumentation and making claims based on evidence.	
Students engage meaningful discourse opportunities every day.	
My lessons/units are centered around real-world issues that are directly related to my students lives or community.	
Students often use media/ technology to connect classroom content to the natural world.	

My current teacher attributes	To what extent do you agree or disagree with this statement. What are your current beliefs or values regarding the statement? Why?
I have to know everything about a particular issue before teaching about it.	
I know everything about climate science.	
I feel comfortable admitting to students when I don't know the answer to their question.	
I am comfortable teaching about open-ended issues where I cannot predict student responses.	
I have to feel like the expert in the room.	

My current teacher attributes	To what extent do you agree or disagree with this statement. What are your current beliefs or values regarding the statement? Why?
I often experience imposter syndrome even for topics that I have strong expertise in.	
I have a strong understanding of the Nature of Science principles.	
I have a strong understanding of the NGSS framework.	

Look over your responses and compare them to the results from the introductory chapter. What shifts in thinking have occurred for you throughout this book? What are some goals that you have now? What were the biggest takeaways from this resource? What support do you still need? What do you feel empowered to do next?

Notes

1 Former governor of California Jerry Brown's remarks were made at the 2019 ECCLPS event hosted at the UCLA Luskin Center.
2 Actor Ricky Gervais' remarks were made in 2017 during The Late Show with host Stephen Colbert.

Collective Voices for Climate Change Education
Emily Courtney (Director of Education Programs at SEI)

Climate Change is a magnifier of social inequity, threatening the safety and livelihoods of global community members who are: most heavily reliant on primary natural resources and least able to move out of harm's way, shelter from extremes, or rebound when knocked over by the consequences of climate change. The good news is that when we **empower** *our students with knowledge of the scientific causes and consequences of climate change and we support them to make this connection, they are able to see the tremendous opportunities that climate change presents for high wage, high growth careers and entrepreneurship that build a life of meaning through high impact service to people and the planet. Students do not have to choose between health, happiness, a good job, and protecting the climate. When we teach our students about climate change, we open the door to an integrated life of meaning through a rewarding, family sustaining green career.*

Shawn Heinrichs (Lead Storyteller at Only One & Co-founder of SeaLegacy)

We must, once and for all, put an end to the debate regarding our species' role in driving the climate crisis, and instead focus all our collective efforts on addressing it head on. The COVID pandemic was a mere ripple in the ocean compared to the tsunami of catastrophic propositions that the runaway warming of our planet is going to result in. Together, we must take a stand for our biosphere, committing to important and decisive measures to massively reduce our dependency on fossil fuels, while also taking immediate action to mitigate the impacts that are already coming our way. We must decide enough is enough and that we will no longer be complicit in global crisis that is driving the collapse of earths biodiversity and now threatening the very survival of our species. **Together we can do this!**

Additional Teacher Resources

Connect with others through the Alliance for Climate
   Education –
   acespace.org/
Explore this website to track water inequities and get involved –
   water.org
Gain access to NNOCCI climate change teaching resources –
   climateinterpreter.org/resources
NGSS Tasks Pre-Screener & Screener –
   www.nextgenscience.org/taskscreener
STEM Teaching Tools Assessment Resources –
   http://stemteaching tools.org/tgs/Assessment
STEM Teaching Tools Equity Resources –
   http://stemteaching tools.org/tgs/Equity
Teaching the United Nations Sustainable Development Goals –
   http://mcic.ca/pdf/SDG_Primer_FINAL.pdf
View Prince Ea's "Sorry" video clip –
   Bit.ly/EASORRY
View Prince Ea's "Man vs Earth" video clip –
   Bit.ly/MANVEARTH

References

Caranto, B. F., & Pitpitunge, A. D. (2015). Students' knowledge on cli-
   mate change: implications on interdisciplinary learning. In *Biology
   Education and Research in a Changing Planet* (pp. 21–30). Singapore:
   Springer.

Carter, B. E., & Wiles, J. R. (2014). Scientific consensus and social controversy: Exploring relationships between students' conceptions of the nature of science, biological evolution, and global climate change. *Evolution: Education and Outreach*, 7(1), 6.

Freire, P. (1972). *Pedagogy of the oppressed.* New York: Herder and Herder.

Hansen, P. J. K. (2010). Knowledge about the greenhouse effect and the effects of the ozone layer among Norwegian pupils finishing compulsory education in 1989, 1993, and 2005 – What now?. *International Journal of Science Education*, 32(3), 397–419.

Hestness, E., McDonald, R. C., Breslyn, W., McGinnis, J. R., & Mouza, C. (2014). Science teacher professional development in climate change education informed by the next generation science standards. *Journal of Geoscience Education*, 62(3), 319–329.

Hodson, D. (2003). Time for action: Science education for an alternative future. *International Journal of Science Education*, 25(6), 645–670.

IPCC (2018). *Global Warming of 1.5°C. An IPCC Special Report on the impacts of global warming of 1.5°C above pre-industrial levels and related global greenhouse gas emission pathways, in the context of strengthening the global response to the threat of climate change, sustainable development, and efforts to eradicate poverty* [Masson-Delmotte, V., P. Zhai, H.-O. Pörtner, D. Roberts, J. Skea, P. R. Shukla, A. Pirani, W. Moufouma-Okia, C. Péan, R. Pidcock, S. Connors, J. B. R. Matthews, Y. Chen, X. Zhou, M. I. Gomis, E. Lonnoy, T. Maycock, M. Tignor, and T. Waterfield (eds.)]. In Press.

Matkins, J. J., & Bell, R. L. (2007). Awakening the scientist inside: Global climate change and the nature of science in an elementary science methods course. *Journal of Science Teacher Education*, 18(2), 137–163.

Somerville, R. C., & Hassol, S. J. (2011). The science of climate change. *Phys. Today*, 64(10), 48.

Appendices

Appendix A   Interactive Science Mapping Tool

Using UC Berkeley's Online Mapping Tool

1. First view allowing for students to select where they started. Click on the bubbles on the right and then fill out the left tabs in detail.
2. The end result will look like the following image. Have students submit their process of science mapping tool as a PowerPoint by selecting the button in the upper right-hand corner with three grey bars.
3. Select the PowerPoint option to download. The PowerPoint file will upload to the desktop. Students can upload to the learning management system for grading, use it as a presentation, or share it for peer feedback.

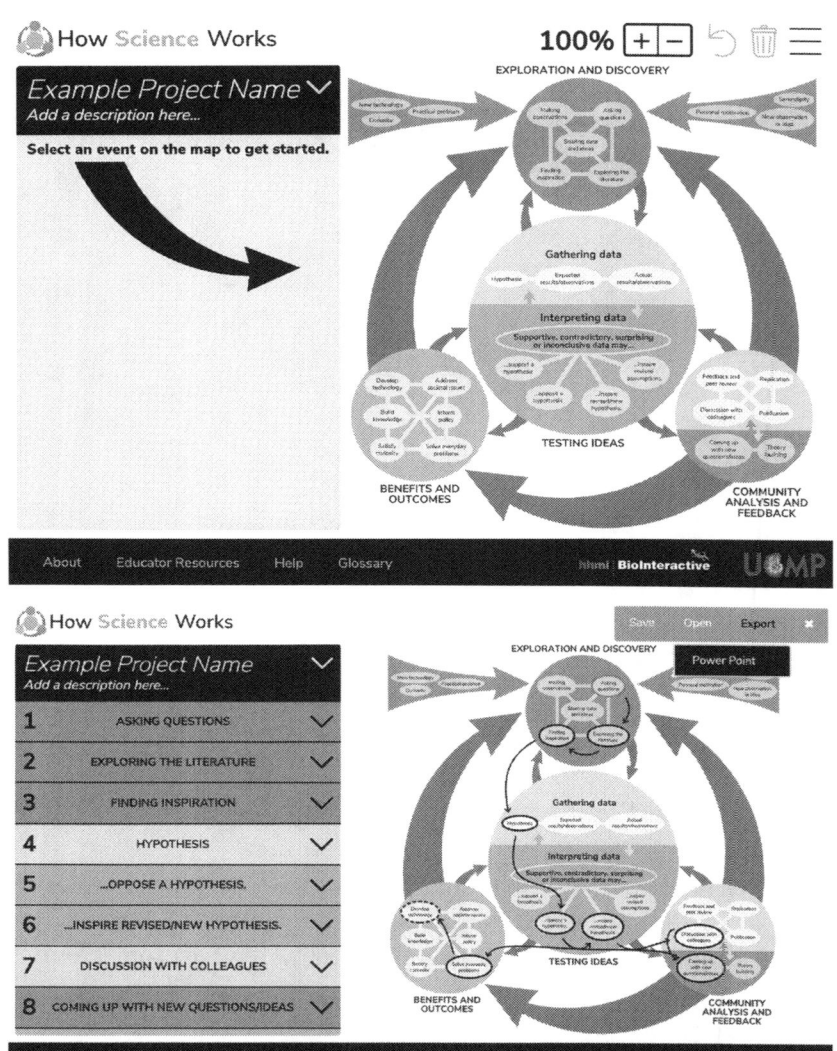

Appendix B Curriculum Map Templates

To access printable versions of these templates to organize the information, please go to www.empoweredscienceteachers.com (under "Resources" tab).

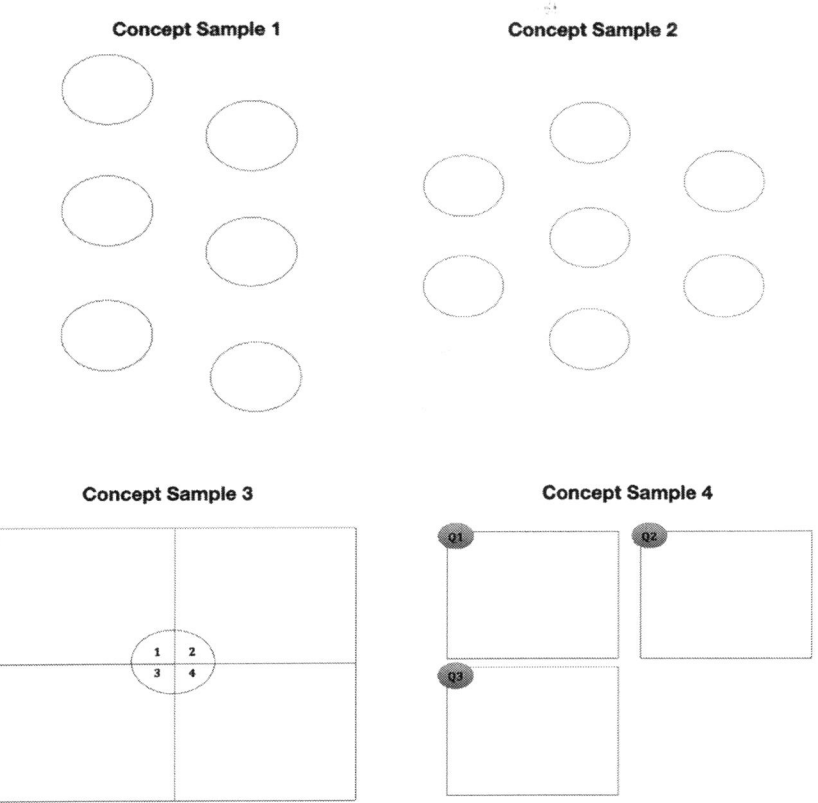

Appendix C Sample Curriculum Maps

The following are curriculum map samples from participants in the climate education program showing their design process over the days (in the order of Biology, Chemistry, and Environmental Science).

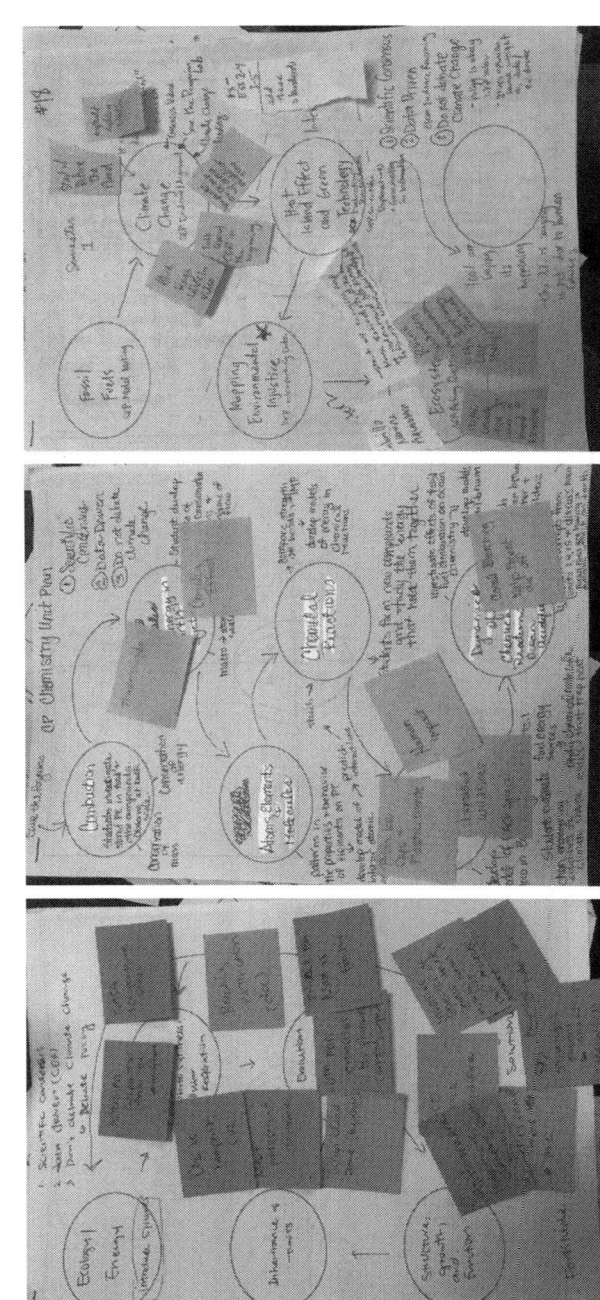

Appendix D   Developing the Driving Question

Developing the Driving Question   ¿ ?

My Original Question

Phenomenon – Something that happens in nature that is not easily explained.

(Write a question about the phenomenon you just saw. What are you curious about?)

Group 1's Question:

Group 2's Question:

Group 3's Question:

Group 4's Question:

Group 5's Question:

Group 6's Question:

Group 7's Question:

Group 8's Question:

Final Anchoring Question:

Appendix E   CER Worksheet

Claim, Evidence, Reasoning (CER) Worksheet

My question:

Driving question: *Write additional questions you have as you learn more in a different color.*

Claim: *Why do you think this is happening?*	
Evidence: *What science content, information, or data supports your claim? What would you use to convince someone that your claim should be adopted? This can be in bullet points.*	
Reasoning: *How does the class information, data, or content support or make you want to adopt, modify, or reject your current claim (Write modified claims under your initial claim)? How does the data or information explain why the phenomenon is happening?*	
Evidence: *What science content, information, or data supports your claim? What would you use to convince someone that your claim should be adopted? This can be in bullet points.*	
Reasoning: *How does the class information, data, or content support or make you want to adopt, modify, or reject your current claim (Write modified claims under your initial claim)? How does the data or information explain why the phenomenon is happening?*	

Appendix F   Mini Lab Portfolio Components

Mini-Lab Portfolio Components

1. Research Topic & Background

What were you looking at specifically for this lab? What are you trying to discover or learn about? Write 2 paragraphs about your lab topic, all the major ideas, any related content that you're applying, state standards that apply, and relate it to this specific lab. What are you research questions? What type of study is this?

2. Materials and Methods

List all materials used and your detailed procedures. This section needs to be explained in your own words to an audience of 7 year-olds (be very detailed about what you did and provide rationale for why you did those things).

3. Creative Title

If someone passed by, this title should captivate them to want to learn more about your topic (Keep in mind your audience and appropriateness).

4. Hypothesis

The hypothesis is written in the format of an "If," "Then," "Because" statement.

5. Abstract

Give a summary of your research topic and lab findings to help others determine if they want to read about your work (Think about what you are trying to find, how you found it, and what you found).

6. Photos and Results

This area should include photos from the experiments and any apparatuses that were used to conduct the experiments (can be drawn nicely if necessary). Include any data tables and graphs analyzing your results. You may attach photos of the lab set up, the process, and the results (before and after). This is where your statistical data should be displayed with mathematical work and results.

7. Discussion and Conclusion

Minimum of two paragraphs discussing your hypothesis (do you accept or reject it based on your results), sources of error that might affect your lab results, and science implications (what can you do with this information? What can be invented? How is this useful? Why should we care?). What major ideas/information will you have a better understanding of now that you've completed the lab? Explain and elaborate on why you think differently now and how. Any works cited can be listed anywhere on the poster.